Human-Nature Interactions in the Anthropocene

Routledge Studies in Environment, Culture, and Society

SERIES EDITORS: BERNHARD GLAESER and HEIKE EGNER

This series opens up a forum for advances in environmental studies relating to society and its social, cultural, and economic underpinnings. The underlying assumption guiding this series is that there is an important, and so far little-explored, interaction between societal as well as cultural givens and the ways in which societies both create and respond to environmental issues. As such, this series encourages the exploration of the links between prevalent practices, beliefs and values, as differentially manifested in diverse societies, and the distinct ways in which those societies confront the environment.

1 **Human-Nature Interactions in the Anthropocene**
Potentials of Social-Ecological Systems Analysis
Edited by Marion Glaser, Gesche Krause, Beate M.W. Ratter and Martin Welp

Human-Nature Interactions in the Anthropocene

Potentials of Social-Ecological Systems Analysis

Edited by Marion Glaser, Gesche Krause, Beate M.W. Ratter and Martin Welp

Associate Editor: Andrew Halliday

Routledge
Taylor & Francis Group
NEW YORK LONDON

First published 2012
by Routledge
711 Third Avenue, New York, NY 10017

Simultaneously published in the UK
by Routledge
2 Park Square, Milton Park, Abingdon, Oxon OX14 4RN

Routledge is an imprint of the Taylor & Francis Group, an informa business

Library of Congress Cataloging-in-Publication Data

Human-nature interactions in the anthropocene : potentials of social-ecological systems analysis / edited by Marion Glaser . . . [et. al.].
p. cm. — (Routledge studies in environment, culture, and society ; v.1)
Includes bibliographical references and index.
1. Nature—Effect of human beings on. 2. Human beings—Effect of environment on. I. Glaser, Marion.
GF75.H88 2012
304.2—dc23
2011040666

ISBN13: 978-0-415-51000-4 (hbk)
ISBN13: 978-0-203-12319-5 (ebk)

Typeset in Sabon by IBT Global.

Printed and bound in the United States of America on acid-free paper by IBT Global.

Contents

PART III
Resilience, Adaptability and Transformability

PART IV
Multi-Agent Modelling and Simulation

PART V
Pathways Towards New Systems Thinking in Human Ecology

Figures

Tables

Foreword

Rik Leemans

Humanity is facing several critical challenges at the beginning of the 21st century. These include food and water security, coping with pandemics, mitigating and adapting to climate change, and slowing the decline in biodiversity. We live in a globalised world in which our most of these problems go well beyond regional and national borders. Understanding, coping with or, if possible, solving these problems requires an in-depth understanding of systemic interactions between the broader marine, terrestrial and atmospheric domains, of dimensions, dynamics and resilience of the systems involved, of tradeoffs, synergies and feedbacks between different parts, and, last but surely not least, of human responses and behaviour. The resulting human–environment interactions within a broader social-ecological system show that simple, deterministic relationships linking environmental stress and social change are inadequate for explaining how humans have historically interacted with nature, and provide misleading guidance for managing our increasingly interconnected world.

The kinds of questions now being asked of the research community have changed over the last decades, reflecting this increasing appreciation of the interconnectedness of the Earth System. As humans tackled scientific questions of societal importance initial investigations were by individuals; later some of the harder questions could be successfully addressed by concentrating on either individual or a small number of the components of the Earth System, although answering such questions required international, interdisciplinary collaboration. For example, understanding the causes and consequences of acid rain required collaboration between atmospheric scientists and terrestrial ecologists. Solving this problem required additional input of economists and political scientists. The emerging interdisciplinary research increased the understanding and helped to develop adequate solutions to many of these environmental problems. Similar research endeavours have to be developed for the current critical challenges, but the disciplinary integration and interdisciplinary understanding must be much further advanced. Additionally, the actual research questions of these societal challenges should be framed and articulated not only by scientists but also more and more with policy-makers and relevant stakeholders, particularly those concerned with adaptation strategies and sustainable development.

The need for an international, multidisciplinary approach to aspects of global environmental change research is not new. For example, the World Climate Research Programme (WCRP) was established in 1980 in response to the need for a framework within which to organise and facilitate international climate research. The International Human Dimensions Programme on Global Environmental Change (IHDP) was started in 1996 to build a support structure for international multidisciplinary research in the social sciences. Each of these international programmes is itself highly international and multidisciplinary in scope and influential. For example, the now widely accepted term 'Anthropocene' stems from one of our workshops (Crutzen 2002). Over the last decade, the Earth System Science Partnership (ESSP) formed to study the Earth System, the ways that it is changing, and the implications for global and regional sustainability in an integrative way (Leemans et al. 2009). The ESSP contributes to this endeavour through a number of activities, including joint projects, capacity building, integrated regional studies, conferences and publications, and contributions to the international science-policy assessments and international conventions.

The ESSP has been, for example, very successful in providing updates on and explaining the trends of the global carbon budget (Le Quere et al. 2009) and developing timely research on food systems (Ericksen et al. 2009). But one of the major struggles remains how to further define, develop and implement integrative research that studies, explains and projects the various interactions within human–environment or social-ecological systems.

This book provides major insights from timely and innovative approaches which bring together findings from natural and social science and especially from the extensive field of systems analysis. It provides a magnificent theoretical and pragmatic overview of models and their applications, theories and conceptual frameworks. In the different chapters issues like complexity, vulnerability and resilience are explained and put in the proper context of these critical challenges. The final synthesis on human ecology and social-ecological systems is incredible. Here inter- and transdisciplinary approaches are comprehensively related to systemic dynamics, human behaviour and the different dimensions and scales. Although the book does not provide solutions to the critical challenges, it outlines the science needed to address these challenges in the near future. That is already a major accomplishment, from which the ESSP certainly will learn.

Wageningen, November 2011
Chair Scientific Committee of the ESSP
Professor in Environmental Systems Analysis, Wageningen University

REFERENCES

Crutzen, PJ (2002) Geology of mankind: The Anthropocene. *Nature* 415: 23

Ericksen, PJ, Ingram, JSI, Liverman, DM (2009) Food security and global environmental change: Emerging challenges. *Environmental Science and Policy* 12: 373–377

Le Quere, C, Raupach, MR, Canadell, JG, Marland, G Bopp, L, Ciais, P, Conway, TJ, Doney, SC, Feely, RA, Foster, P, Friedlingstein, P, Gurney, K, Houghton, RA, House, JI, Huntingford, C, Levy, PE, Lomas, MR, Majkut, J, Metzl, N, Ometto, JP, Peters, GP, Prentice, IC, Randerson, JT, Running, SW, Sarmiento, JL, Schuster, U, Sitch, S, Takahashi, T, Viovy, N, van der Werf, GR, Woodward, I (2009) Trends in the sources and sinks of carbon dioxide. *Nature Geoscience* 2: 831–836

Leemans, R, Asrar, G, Busalacchi, A, Canadell, J, Ingram, J, Larigauderie, A, Mooney, H, Nobre, C, Patwardhan, A, Rice, M, Schmidt, F, Seitzinger, S, Virji, H, Vörösmarty, C, Young, C (2009) Developing a common strategy for integrative global change research and outreach: The Earth System Science Partnership (ESSP). *Current Opinion in Environmental Sustainability* 1: 4–13

Preface

The German Society for Human Ecology (DGH) and Land-Ocean Interactions in the Coastal Zone (LOICZ)

Bernhard Glaeser

Human Ecology is a young discipline. Its scientific object is the interrelations and interactions between society, humans and the environment. Its core is a holistic approach that includes physical, socio-cultural, economic and political aspects. The term 'human ecology' was coined by the Chicago School of sociology in the 1920s. It has spread ever since as a research perspective in the natural, social and planning sciences. In some countries, university chairs were established.

The German Society for Human Ecology (DGH) was founded in 1975. In its earlier years, it was largely influenced by social medicine. Later, it expanded thematically into other fields of science and policy. Today, DGH is a forum which hosts and assembles experts pertaining to all environment-oriented disciplines. Further details, including workshops, themes and publications, can be found at the homepage (www.dg-humanoekologie.de).

Ever since 1989, the DGH annual meetings have been devoted to the interdisciplinary discourse in focal topics relevant to varying environmental issues. This resulted in a series of book publications published in German by Westdeutscher Verlag. In 2000, DGH decided on a new start with Oekom Publishers and the new series Edition Humanökologie.

This present book addresses the international community. A better understanding of human–nature dynamics is the core aim of this volume. In the age of the Anthropocene, human-induced changes surpass those deriving from non-anthropogenic, natural dynamics. The question is how social-ecological systems analysis can contribute. For that, complementarities and synergies between human ecology, systems theory and the various branches of social-ecological analysis are explored. This is another attempt in inter- and transdisciplinary research to link different knowledge systems with innovative methodological approaches.

The Sommerhausen 2008 symposium was co-organised and co-sponsored by DGH and Land-Ocean Interactions in the Coastal Zone (LOICZ), a joint core project of the International Geosphere-Biosphere Programme (IGBP) and the International Human Dimensions Programme on Global Environmental Change (IHDP). Social-ecological systems analysis on global coasts is one of three priority topics of LOICZ as well as an eminent theme in human ecology and thus a good reason for close cooperation between DGH and LOICZ.

Berlin, December 2011
DGH President and LOICZ Scientific Steering Committee Member

Acknowledgments

The editors wish to particularly thank the five indefatigable, well-humoured assistants who helped this book along its various stages: Kim Smyla, who carried the administrative load of organising the Sommerhausen Symposium with humour and skill; Stefanie Dreger and Barbara Kummetz, who cheerfully shared the initial book formatting and reference sorting process; and eventually Jana Bunk and Nadine Grund, who shared the "final lap" towards completing the manuscript. Andrew Halliday became so important, not only for the English language editing he undertook on the entire—often somewhat "Germlish"—manuscript, but also as a co-author in Chapters 4 and 10 that the original editors agreed to name him Associate Editor. Special thanks are due for his tenacity, attention to detail and concise critiques and analytical inputs throughout. Five (!) anonymous reviewers provided valuable (albeit sometimes starkly contrasting) recommendations. We are grateful for their time and inputs.

Part I

Introduction

1 New Approaches to the Analysis of Human–Nature Relations

Marion Glaser, Beate M.W. Ratter, Gesche Krause and Martin Welp

In interaction with nature, humans shape the conditions for their own survival, quality of life and future options. Humans are also endowed with reflexivity, that is, with the capacity to deliberate on the consequences of their actions and to decide to alter behaviour in order to produce more desirable outcomes. This gives humans exceptional power, compared to all other species, to transform their environment. Human reflexivity is inextricably linked to analysis. The analysis of human–nature relations in our contemporary age of the Anthropocene (Crutzen 2002), a new geological epoch, where human activity is the main driver of planetary evolution (3rd Nobel Laureate Symposium on Global Sustainability 2011), is the theme of this book. Specifically, the book explores the potential of transdisciplinary approaches for the analysis of human–nature dynamics. We argue that a reorganisation of scientific endeavour along these lines is urgently needed in order to address threats to the sustainability of global ecosystems, and thus to human life on earth.

This book, and the German Society for Human Ecology (DGH) Symposium in Sommerhausen, Main, Germany (May 28–31, 2008) which gave rise to it, were sponsored by the Land–Ocean Interactions in the Coastal Zone (LOICZ) programme through Priority Topic 1 (Social-Ecological Systems Analysis). They continue the work of previous international symposia and workshops, also organised under the LOICZ umbrella in Beijing and Rio de Janeiro in 2007. These brought together academics from multiple disciplines and continents to discuss the theme of 'social-ecological systems analysis in the coastal zone'. The central questions addressed in these symposia were as follows: How do you define a social-ecological system (SES) of interest? For a given SES, what are the drivers of change? What range of future scenarios can be envisaged? How can sources of resilience, vulnerability, adaptability and transformability be identified? What are the options for steering the system towards more desirable future trajectories and how can these be agreed upon? As a frame of reference for these discussions, participants at the Beijing and Rio de Janeiro symposia were asked to critically refer to the terminology and definitions presented in Box 1.1.

The SES Conceptual Framework

A social-ecological system is a complex, adaptive system consisting of a bio-geophysical unit and its associated social actors and institutions. The spatial or functional boundaries of the system delimit a particular ecosystem and its problem context.

Resilience is the capacity of a system to handle whatever the future brings without being altered in undesirable ways. Resilience is necessary for a sustainable future and is rooted in the in self-reinforcing dynamics of natural systems that prevent shifts into undesirable directions.

Vulnerability refers to the degree to which a system is exposed to undesirable consequences of change.

Adaptive capacity and *transformability* refer to system's ability to change, in the former case to maintain desirable features of the system, and in the latter to eliminate undesirable ones.

Emergence is generated through the capacity of systems to self-organise without external direction.

Box 1.1 Key terms for social-ecological systems analysis, with definitions (slightly adapted) used at Social-ecological Systems Symposia in Beijing 2007 and Rio de Janeiro, 2007 (see Glaeser et al. 2009).

Some main conclusions of the Beijing and Rio de Janeiro symposia were the following:

- SES terminology needs to be interpreted and operationalised taking account of context-specific social and ecological conditions. This includes a methodologically adequate treatment of local knowledge and priorities, which at present is rare.
- SES analysis cannot incorporate all the complexity of real-world interactions between humans and their environment. In order to reduce this complexity in a societally relevant manner, SES analysis should focus on addressing specific problems in determined geographical areas.
- Interdisciplinary ventures should be planned with a balanced distribution between the natural and social sciences in terms of senior positions, numbers of staff, budgets and planning responsibilities and rights. Ideally, programme administrations and funding should come from sources oriented towards interdisciplinary knowledge generation (Glaeser et al. 2009).

The Sommerhausen Symposium took up some of these challenges by addressing the questions listed in Box 1.2.

1. Which system concepts best describe which types of social-ecological system?
2. What are the differences, complementarities and redundancies between the concepts of system resilience, adaptability and transformability?
3. How do we explain social-ecological emergence? To what extent can agent-based models simulate emergent change at higher system levels, and what are their limitations? Which alternative approaches exist?
4. What are the experiences with stakeholder participation in SES analysis?
5. How can SES analysis contribute to social learning? What is required from science, higher education, schools and further education? Which tools exist?
6. How can SES analysis improve the sustainability prospects of the human–nature connection?
7. How can SES analysis contribute to the identification of feasible pathways towards sustainability?

Box 1.2 Questions originally presented at the 2008 Sommerhausen Symposium (Glaser et al. 2008).

The Sommerhausen Symposium and, even more so, the process of developing this book evolved some unexpected emergent properties, connected to the way contributors reacted to our original questions but also in response to time lags, surprises and synergies. Thus some of the questions we originally prepared (Glaser et al. 2008) evolved into new ones, some were rephrased, and others relegated to marginality. The final chapter of this book thus summarises our conclusions on the somewhat changed but closely related set of six questions listed in Box 1.3.

1. What is required from sustainability-oriented SES analysis?
2. Who needs complex systems thinking, and what for?
3. Which SES concepts best describe which type of system?
4. How can SES analysis contribute to the identification of feasible pathways towards sustainability?
5. What are the experiences with stakeholder participation in SES analysis?
6. What is needed to view the earth as a system and to pre-empt dangerous change?

Box 1.3 Major questions addressed in the book.

THE ANTHROPOCENE

The questions the authors address in this book respond to the new conditions for the analysis of human–nature relations in the Anthropocene. Our overarching aim is to develop a systems-oriented human ecology to contribute to sustainable future human–nature relations from the local to the global level.

Humankind has now set foot in most places on earth. Over a decade ago, Vitousek et al. (1997) documented the increasing proportion of the biosphere appropriated by humans. The term 'Anthropocene' was then coined by Nobel Prize–winner Paul Crutzen (Crutzen and Stoermer 2000; Crutzen 2002). Over the past 300 years, human activities have become an increasingly significant force affecting the evolution of the earth system in its geological, hydrological, bio-geochemical and atmospheric realms. The Millennium Ecosystem Assessment found that, in the past 50 years, ecosystems are changing more rapidly and extensively as a result of human activity than ever before in the history of the biosphere. That this is occurring largely to meet rapidly growing human demands for food, fresh water, timber, fibre and fuel (Millennium Ecosystem Assessment 2005) demonstrates that social processes are driving these changes and highlights the increasing intensity and extent of human–nature feedbacks in the earth system. In response to these challenges, Schellnhuber et al. (2010) point the way towards a convergence between earth system analysis and global sustainability analysis. This book is located within this ongoing process and is intended to advance it.

OVERCOMING BARRIERS BETWEEN DISCIPLINES

How has science, our major and most influential mode of knowledge generation, dealt with the challenges of the Anthropocene so far?

A number of specialist scientific disciplines, including environmental sociology, ethnobiology, environmental anthropology, integrative geography, political ecology and even ecological economics, have made the analysis of human–nature feedbacks a central concern. However, for the majority of academic disciplines, these sub- and interdisciplinary approaches to the study of human–nature relations has remained a sideline, dwarfed by the central themes and mainstream concerns of the respective main disciplines. This marginalisation of the study of human–nature relations at the fringes of its various academic host disciplines is currently prejudicing global efforts to address sustainability problems in human–nature relations.

There is thus a need to improve the positioning of fields of study which deal with human–nature relations within their home disciplines. Communication across disciplines on the subject of human–nature dynamics is also crucial. Systems analysis or modelling provides heuristics and tools for bridging across disciplines; however, there are many stumbling blocks. Disciplines differ in their basic paradigms, assumptions and priorities, disciplinary language, modes of communication and methodological approaches. Not least, qualitatively and quantitatively oriented researches have distinctly different ways of assessing data quality. For instance, although it also generates some divisions within social science disciplines, the debate around the respective merits of statistical significance versus qualitative depth as research-quality criteria is the major dividing line between the natural and the social sciences. Disciplines also differ in their willingness to incorporate different knowledge bases (e.g., stakeholder knowledge, local knowledge and traditional knowledge) into the knowledge generation process. Interdisciplinary and transdisciplinary studies play a pivotal role in bringing these different realms together, bridging different perspectives, promoting dialogue around objectives, and apportioning tasks.

Despite these difficulties, there are a number of flagship initiatives and pilot projects which show that working across the disciplinary divide is possible if seriously intended and properly managed. These examples are explicitly oriented towards holistic, systems-based research for the assessment of policy options to support sustainable management. For instance, SPICOSA, an integrated project funded by the Sixth EU Framework Programme since 2007, brings together 54 European coastal management institutions. The programme is based on an adaptation of the systems approach that incorporates the ecological, social and economic dimensions of coastal zones. The approach is "to provide a cumulative knowledge base for wider and more complex applications, and to contribute to a growing understanding of the options and actions necessary for the transition to sustainable development" (see www.spicosa.eu/). In the tropical coastal and marine realm, the German Leibniz Center for Tropical Marine Ecology (www.zmt-bremen.de) conducts explicitly interdisciplinary research in which natural and social scientists cooperate with modellers, ecosystem users and decision-makers to generate knowledge for a sustainability-oriented governance and management of coastal and marine areas. Also in Germany, sustainability at the interfaces between nature and society is the focus of the Frankfurt Institute for Social-Ecological Research (ISOE) (www.isoe.de; see Becker and Hummel, both this volume). The Stockholm Resilience Centre advances transdisciplinary research for the governance of SES with a special emphasis on resilience, understood as the ability to deal with change and continue to develop. Established in 2006, the centre aims at combining new forms of cooperation with a holistic perspective, in order to generate insights to strengthen societies' and ecosystems' capacities to meet global needs (see www.stockholmresilience.org).

MAJOR LINES OF ARGUMENT

These diverse initiatives share the recognition that mechanistic, linear thinking is inadequate for the complex contemporary challenges associated with human–nature interactions. The authors of this book concur with this view. We conceptualise the earth and its people as an increasingly integrated, global SES that is characterised by surprise, non-linear behaviour and unpredictable changes of trajectory. We offer an analysis of how SES analysis can contribute towards a better understanding of this complex system to help to manoeuvre human–nature relations onto feasible and sustainable pathways. The book combines theoretical debate and a case study presentation in five parts, which include three thematic sections.

Part I (this chapter and the following) outlines the aims and content of the book and discusses the nature of social-ecological thinking and its relevance for social learning for sustainability. The book's first thematic section (Part II) on "Social-Ecological Systems (SES) and Complexity" outlines different types of human–nature systems and traces pathways towards understanding them. Part III focuses on "Resilience, Adaptability and Transformability" and delves further into the impacts of changing human–nature relations on social and ecological sustainability. Part IV examines multi-agent modelling as a transdisciplinary tool for analysing human–nature relations. The section introduces a recent innovative modelling case study in this field.

The concluding chapter in Part V assesses different approaches to human–nature analysis in the Anthropocene. It argues that a range of approaches are 'fit for purpose' in different circumstances, while a comprehensive transdisciplinary approach that incorporates a full range of ecological and social sustainability concerns into global earth systems science is yet to be achieved.

PART I: INTRODUCTION

Following this introductory chapter, the outcomes of an interactive exercise conducted during the Sommerhausen Symposium in May 2008 on 'systems thinking in society' provide the point of departure for the exploration of SES in this book. Gesche Krause and Martin Welp conducted this event, the "World Systems Café", in which participants debated the nature of SES thinking its relevance for social learning. Questions and starting points for the discussion at the symposium were generated here. This collective experience underlined the wide range of views and disciplines related to human–nature interactions among participants at the symposium. Chapter 2 describes the discussions and puts their outcomes into a theoretical perspective.

Participants at the World Systems Café addressed four questions in rotating groups:

1. Who needs an understanding of systems thinking?
2. What are the arenas where people can learn systems thinking and put knowledge into action?

3. What tools, methods and approaches can scientists provide to bring systems thinking into society?
4. What do you *not* perceive of the world when you apply systems thinking?

Krause and Welp present the conclusions of the discussions in the form of a number of central statements proposed by World System Café participants. It is proposed that that the application of systems concepts, such as stock and flows, non-linear behaviour, and feedback loops, is essential in order to comprehend and respond to global environmental issues such as climate change. Currently, however, most people are not familiar with these concepts. However, everybody should and can become a systems thinker. Systems thinking can take place in multiple arenas that serve different purposes and suit different stakeholders, ranging from computer games, to neighbourhood planning exercises and formal education at all levels. Together these can help make systems science accessible across social and economic divides. Krause and Welp also caution against an over-emphasis on technical issues and stating that systems thinking needs to be complemented by an "emotional driver to reach people".

PART II: SOCIAL-ECOLOGICAL SYSTEMS AND COMPLEXITY

The three contributions in the first thematic section trace different pathways towards defining SES and understanding social-ecological change.

In Chapter 3, Egon Becker, a founding father of the new Frankfurt school of Social Ecology, focuses the analysis on *societal relations to nature.* In his presentation to the symposium, Becker characterised the world as a "crisis-ridden, self-organising complex system" of which humanity is an integral part and increasingly powerful driver. This requires a "new world view" and is the point of departure for conceptualising the *provisioning functions of nature* as the focus of Social Ecology. Becker presents the SES concept as a 'boundary object' which is situated at the interface between the different fields of any enquiry. For the purpose of scientific enquiry, this boundary object is transformed into an 'epistemic object', that is, an object of study, whose system properties are formally defined to respond to the research interests of the scientists involved. Becker explicitly contrasts such epistemic objects with 'real-life' social-ecological dynamics.

In Chapter 4, Felix Tretter and Andrew Halliday discuss the usefulness of modelling as a heuristic approach for exploring human–nature relations. They emphasise the importance of clearly defining conceptual frameworks, objectives, and spatial, temporal and thematic boundaries in order to provide a secure foundation for modelling the complexity of SES. The iterative process of building up the model is conceptualised as an *epistemic cycle* which enables feedback between qualitative and quantitative analytical stages. The authors draw on illustrative examples to outline a workable generic approach for moving from the qualitative, conceptual, and often exploratory stages of 'verbal' models to an explicitly quantitative formulation of formal mathematical models which can be used for the computer-based simulation of system behaviour.

In Chapter 5, Beate Ratter outlines the need for definitional and conceptual precision in using the terms of 'complexity' and 'emergence' for analysis of the human–nature interface. Pointing out the difference between structural and behavioural complexity, Ratter shows that complex system *behaviour* is distinct from and not necessarily associated with a complex system *structure*. She argues that behavioural complexity emerges from the relations between system elements rather than from system structure. Ratter perceives nature and society as driven by multi-level interactions of agents which generate emergent structures. She distinguishes between emergent patterns and other possible higher-level system outcomes. Emergence is created "not by a single event or rule" but by "things of an unlike kind", that is, by multiple actors and rules in the context of non-linearity and with the constant imminence of surprise.

PART III: RESILIENCE, ADAPTABILITY AND TRANSFORMABILITY

The book's second thematic section delves further into the exploration of change in human–nature relations by exploring the resilience, adaptability and transformability of social-ecological relations under pressure.

Klaus Eisenack's Chapter 6 is a contribution to the discussion of micro–macro interactions that is assuming increasing importance in the context of climate change and other global changes affecting the earth system. Eisenack and his colleagues at the Potsdam Institute for Climate Impact Research (PIK) have proposed a way to bridge the gaps between local, regional and global sustainability concerns. The concept of *archetypes*, defined as clusters of dynamic variables which generate symptomatic patterns in human–nature relations, is central in this approach. In this chapter, Eisenack's application of the archetype approach to the question of adaptation to climate change draws on the *action theory of adaptation* to identify clusters of mutually reinforcing trends which translate into either desirable or undesirable system dynamics and states (respectively denominated *paradigms and syndromes)*. This is aggregated into a preliminary list of *archetypal barriers to successful adaptation*. This approach creates a clear interface between social-ecological studies with the analysis of policy processes and governance structures.

The explicit focus on basic human needs characteristic of the new Frankfurt School of Social Ecology is evident in Chapter 7. Here, Diana Hummel demonstrates the potential of social-ecological analysis for generating insights into the dynamic relations between human individuals, nature and society. In her presentation at the Sommerhausen Symposium, Hummel explained that her research group define *supply (or provisioning) systems* as "structures to provide the population with basic goods and services which constitute specific societal relations to nature". In Chapter 7 of this book, she emphasises that supply systems respond to human well-being concerns in ways that are dependent on and embedded within nature. She challenges the view that absolute population numbers are the only demographic feature to influence societal access to ecosystem services. With a review of major debates in the fields of population studies, environment and development, and drawing on a case study from urban Ghana, she argues that it is not absolute population numbers but the adaptive capacity of the respective *supply or provisioning systems* which determine the satisfaction of nature-dependent basic human needs. Problems such as hunger and drinking water scarcity are, in this view, a result of inadequate adaptive capacity rather than of absolute supply and demand mismatches.

PART IV: MULTI-AGENT MODELLING AND SIMULATION

This third thematic section specifically explores the potential of multi-agent modelling for the analysis of change in human–nature relations.

In Chapter 8, Marion Glaser draws attention to the prevailing scarcity of methods for integrated social-ecological analysis. She argues that new methods for SES analysis need to deliver on at least three counts: (1) by producing relevant new knowledge on SES, (2) by providing a methodology for the analysis of complexity and emergence, and (3) by facilitating the participation of ecosystem users in systems analysis. The chapter reviews the procedural steps in multi-agent modelling and explores the achievements and potential of multi-agent modelling to meet these requirements. In a discussion of seven exemplary interdisciplinary models, Glaser finds evidence of the growing capacity of multi-agent models to reconstruct patterns of past system behaviour and to simulate plausible future scenarios. She emphasises the potential for linking knowledge systems through participatory multi-agent modelling. She concludes by discussing areas where more work is still needed, including the construction of *interactive models which incorporate human and non-human living agents*, the development of *improved procedures for the participatory ground-truthing* of model outputs, and last, but not least in the context of the Anthropocene, the *systematic relating of agent-based models to earth system level analysis.*

In Chapter 9, Martin Wildenberg and Simron Jit Singh present a prototype multi-level model of the SES of Kamorta Island in the Nicobar Island Archipelago. The focus is on changes brought about by the devastating Tsunami that hit the islands in 2005, and the equally profound impacts of the post-tsunami reconstruction effort. Wildenberg and Singh address the challenges of modelling the "natural and cultural spheres of causation" in this SES by combining an agent-based model with input-output models of selected socio-economic and ecosystem system components. To simulate outputs under a range of different conditions, the interaction of these modules is modelled mathematically. The results provide insights into the co-evolution of lifestyle preferences, land use, and vegetation cover in the research area in the post-tsunami period. A novel feature of the model is that ecosystem users are able to participate in the modelling process by using slide bars to vary conditions and observe how alternative policy choices can be expected to affect system outcomes.

PART V: PATHWAYS TOWARDS NEW SYSTEM THINKING IN HUMAN ECOLOGY

Chapter 10 reviews the principal conclusions of previous chapters in the light of recent developments in earth system science and global sustainability analysis. We revisit some of the key questions initially posed to the Sommerhausen Symposium (see Box 1.2) and provide some answers to the key questions on SES analysis that emerged during the preparation of this book (see Box 1.3). Future directions for the analysis of human–nature interactions in the Anthropocene era are discussed. A key conclusion is that human well-being concerns need to be more explicitly incorporated into SES analysis at all scales in order to increase the relevance of scientific knowledge generation to the multiple decision-makers whose actions will determine the future of our planet.

REFERENCES

3rd Nobel Laureate Symposium on Global Sustainability (2011) Executive Summary: Transforming the World in an Era of Global Change. May 16–19, 2011, Stockholm. Available at http://globalsymposium2011.org/wp-content/uploads/2011/05/resilience_summaryXlow2.pdf

Crutzen, PJ (2002) Geology of mankind: The Anthropocene. Nature 415: 23

Crutzen, PJ, Stoermer, EF (2000) The "Anthropocene". Global Change Newsletter 41: 17–18

Glaeser, B, Bruckmeier, K, Glaser, M, Krause, G (2009) Social-ecological systems analysis in coastal and marine areas: A path toward integration of interdisciplinary knowledge. In: Lopes, P, Begossi, A (eds) *Current Trends in Human Ecology*. Cambridge Scholars Publishing, Cambridge, pp 183–203

Glaser, M, Krause, G, Ratter, B, Welp, M (2008) Human–nature interaction in the Anthropocene: Potential of social-ecological systems analysis. GAIA 17: 77–80

Millennium Ecosystem Assessment (2005) *Ecosystems and Human Well-being: General Synthesis*. Island Press, Washington, D.C.

Schellnhuber, HJ, Molina, M, Stern, N, Huber, V, Kadner, S (2010) *Global Sustainability: A Nobel Cause*. Cambridge University Press, Cambridge

Vitousek, PM, Mooney, HA, Lubchenco, J, Melillo, JM (1997) Human domination of Earth's ecosystems. Science 277: 494–499

2 Systems Thinking and Social Learning for Sustainability

Gesche Krause and Martin Welp

INTRODUCTION

Publications and conferences over recent decades highlight the breadth of disciplinary interest and the wide range of different perspectives in the field of systems thinking (e.g., Checkland 1984; Midgley 2003; Siebenhüner 2004; Renesch 2008). This is reflected also in different approaches to social-ecological systems (SES) thinking discussed elsewhere in this book. In particular, studies of natural and social systems have developed in radically different directions. Current work on social and natural systems derives from the two strands of systems thinking, with two more or less independent points of origin. This provides exciting opportunities for new and emergent ways of thinking that go beyond traditional disciplinary boundaries. Engaging with other fields of research can however be a time-consuming process. It is hampered by the difficulty of fully endorsing the rationality behind the strands of systems thinking relating to disciplines other than our own; since the way we understand systems is also a reflection of how we, as individual scholars, have been exposed through social learning to systems thinking. The study of geology, biology or sociology, for example, provides different personal mental models as starting points for systems thinking. The Nuffield Review found that students going to university increasingly struggle with work that requires them to think independently or to make connections between different, often narrowly defined areas of study (Caulkin 2006).

The objective of this chapter is to explore the application of systems thinking within broader social learning, that is, the learning of whole societies as a common endeavour (Kates et al. 2001; Siebenhüner 2004). Our premise is that sustainability would benefit from a better understanding of systems thinking among ordinary citizens. It is not sufficient for only experts to be knowledgeable about systems thinking. The public needs systems thinking in order to understand processes that take place in our economies, environment and societies. This involves time lags, non-linear behaviour and feedback loops, and other patterns of behaviour that are typical of complex systems. If long-term political measures, for example to mitigate climate change, are to be legitimised in a democratic decision-making process, citizens need a level of systems understanding that is appropriate to the situation. What this appropriate level implies and what kind of approaches to learn systems thinking are useful in practice is one of the core issues to be discussed in the following contribution. The paper was inspired by a World Café exercise with academics engaged in systems analysis and presents some results of this exercise.

THE *WORLD SYSTEMS CAFÉ* EXERCISE

The organisers of the Annual Conference of the German Society of Human Ecology in Sommerhausen (May 2008) chose to hold a World Café exercise after having had good experiences with this method during earlier conferences and workshops. The World Café is a large group method, which contains a sequence of discussions at tables with four to seven people seated at each table (Brown and Isaacs 2005). Such sessions can last from a couple of hours up to one to two days. In our case, the session took place during one afternoon and was continued in the following morning session. One person per table acted as a host. The other participants were able to move between tables between sessions, each of which revolved around a single question which was discussed for between 20 and 30 minutes by the attendants at the respective table (Figure 2.1). Large sheets of paper were supplied to

Figure 2.1 World Systems Café exercise in Sommerhausen 2008.

each table so that people could draw and write down their ideas. Due to the topical focus on systems thinking, the session was given the name 'World Systems Café'.

In our World Systems Café sessions, we collected thoughts and personal experiences of experts on the question of whether there is a need to root knowledge of systems thinking into society at large and, if so, how to go about it. The central question of the exercise concerned the main arenas and tools available for achieving this in practice. The following questions were introduced by the World Systems Café moderators and addressed by the participants in four subsequent discussion rounds:

- Who needs an understanding of systems thinking?
- Where can people learn systems thinking and put knowledge of it into action?
- Which tools, methods and approaches can scientists provide to bring systems thinking into society?
- What do you not perceive of the world when you apply systems thinking?

The World Systems Café approach brought out participants' reflections on their personal experience and their opinions on the questions being discussed. In the following sections we present, for each of the four questions, several provocative statements that emerged in the discussions. A consensus view on some of these statements was reached fairly quickly, while diverging views remained on others. We provide a short summary of arguments from the group work exercise. We then look into the body of relevant literature to see whether and how these assumptions, which emerged from a group of experts in the field of systems analysis, are discussed. Even though most of the participants were researchers in the field of human ecology, the discussions revealed that there are many different views on systems thinking, thus underlining the wide breadth of different scientific views and disciplines involved in the research field of human–nature relationships.

WHO NEEDS AN UNDERSTANDING OF SYSTEMS THINKING?

Summary of Group Discourse

The first question provoked an extensive discussion on 'What does systems thinking consist of?' at all tables. For example, is it sufficient to state that systems thinking can be linear or non-linear? This question was left open, but different opinions were reflected in the diverging views expressed in the three key statements that follow. The statements were extracted from the discussions and presentations on the first question, 'Who needs an understanding of systems thinking?'

Statement 1:
Every human being is a systems thinker.

Several groups argued that human survival is rooted in the notion of systems thinking, since systems thinking orders the world in a certain way. Thus, a formal declaration of systems thinking is not necessary, because it is implicit knowledge that is available to all mankind, that is, everybody has it but nobody knows it. Thus, what we need is a better understanding of the issues of complexity, uncertainty and connectedness, in systems thinking terms, and we need to learn to ask critical questions. This holds especially true for decision-makers.

Statement 2:
Everybody needs to become a systems thinker.

Other groups expressed the view that systems thinking provides us with the means to see more possibilities and to translate them into action. Everybody should have an understanding of systems thinking, but not necessarily to the same degree. Aristotle's phrase that "*the whole is more than the sum of its parts*" was cited to underline this viewpoint, implying that the total social understanding of systems thinking is greater than the sum of individual ones. It was stressed that putting knowledge into action would require changes to some (or most) arenas of learning, that is, schools or universities. However, some argued, if one started early with children, knowledge of systems thinking would be in danger of being instrumentalised. If real-world problems such as the impacts of climate change were addressed in schools in early classes there would be a risk of transmitting overly negative messages, leading to frustration. It was further argued that we would need to think of more positive ways of communicating systems ideas, drawing on a wider range of examples of applications of systems thinking in different fields. For example, the framing of the climate issue matters a lot, that is, unavoidable catastrophe or solvable challenge. The motivation to cooperate and to find practical solutions that actually work in the societal realm are key issues yet unresolved, especially with respect to sustainability and the management of global commons. One group pointed out that the arenas to put knowledge into action could be situated on different scales.

Statement 3:
Everybody can learn systems thinking.

Some participants of the World Systems Café argued that, for instance, students of environmental management ought to and can learn systems thinking. Hereby the discussions focused mainly on the use of modelling tools as one promising path employed for systems thinking for students at university level. It was stated that at least from master's level onward, all students should have necessary analytical skills to undertake simple modelling exercises using software tools like Vensim or Stella. However, what about school children? At what stages should elements of systems thinking be included in school curricula? Although many school children are experienced in the use of computers, computer-based modelling is not something that everybody is comfortable with. Are there system-learning methods apart from computational modelling that can be applied in schools? It was argued that the heterogeneous backgrounds of different scientific fields act as a major constraint to fostering systems thinking at an early stage in schools.

Discussion

"If someone were to analyse current notions and fashionable catchwords, he would find 'systems' high on the list". This statement by von Bertalanffy in 1968 is more valid today than ever before. By and large, systems thinking is just one way of looking at the world around us, and could be defined crudely as the science of systems. Systems thinking, under the name of General Systems Theory (GST), was first developed during the 1930s and 1940s in response to the search for universal principles applicable to more than one of the traditional departments of knowledge (Arias Schreiber and Glaser 2008). According to the scholars of modern systems thinking (e.g., von Bertalanffy 1968; Laszlo 1972), a system can be understood as a defined set of objects together with the relationships between these objects and between their attributes (Hall and Fagan 1956). This view entails a set of characteristics and concepts, such as wholeness and sum, environment, hierarchical order, and so on, which are common to all systems. Although the founders of GST hoped that it would unify science, systems thinking and practice has branched out into numerous sub-disciplines in a similar way to traditional science. In particular, studies of what are seen as natural and social systems have diverged and developed in radically different directions. Thus we can say without much exaggeration that systems practitioners currently working in social and natural sciences speak different languages and inhabit different worlds. How many readers of *Ecology and Society*, for example, also subscribe to *Systems Research and Behavioural Science*? Thus, forming linkages between social and natural systems-thinking scholars presents both conceptual and cultural challenges.

In both the natural and social sciences, the growing interest in systems can be linked to a number of parallel research trends, visible over recent years, which all involved the call for paradigm shifts towards more interdisciplinarity. This is seen as a prerequisite for achieving sustainable management of natural resources, since the problems that need to be tackled are too broad or complex to be resolved by a single discipline (Arias Schreiber and Glaser 2008). That the same applies to social problems was underlined by Sterman (2002), who raises the point that it takes a village to raise a child and argues that this is a systems perspective in its clearest form. The community is needed because no one individual could possibly provide for the child all the knowledge and experience needed to nurture a well-adjusted adult. Similarly, society is needed to nurture a healthy environment (Keen et al. 2005). This view was reinforced by Laszlo (1972) who stated that "if we are to understand what we are and what we are faced with in the social and the natural world, evolving a general theory of systems is imperative".

However, if we go beyond the confines of the academic world, perspectives on systems thinking become even more diverse. For instance, Sterman (1994) conducted experiments with MIT undergraduates and demonstrated that students have serious problems in engaging even with relatively simple examples of systems thinking. If university undergraduates have serious difficulties in understanding and interpreting, for example, simple models of stocks and flows that can be regarded as one very simplistic expression of systems thinking, one can assume that ordinary citizens will face similar problems.

Using the climate change issue as an example of a topical global change problem requiring urgent action, social science studies of lay people's understanding of climate change suggest that most people perceive it as a linear system (Sterman 2008). This way of thinking is reflected in the common assumption that when emissions go down, the temperature will go down too. This public understanding is, however, misleading since it ignores the basic functioning of stocks and flows and system inertia, not even mentioning non-linearities and critical thresholds (e.g., tipping points). The consequences of such misunderstandings can be illustrated by the reaction to results of recent studies suggesting that it is already inevitable that mean global temperatures will rise by at least 2.4 degrees Celsius (Ramanathan 2008). This finding, if correct, would imply that the 2 degree target, defended by many scientists and policy-makers, is already unachievable. However, this cited study hardly gained any attention in the public media. When it comes to creating political momentum, as is required for the acceptance for carbon taxes, economic incentives for renewables, or carbon trading mechanisms, such misperceptions may lead to a severe underestimation of the urgency of the actions required.

Thus, not only academics, but all social groups within a society (e.g., politicians, decision-makers, ordinary citizens, children, etc.) need a better understanding of the linkages between social and natural systems. This has become one of major themes in the debate about sustainability. To date, systems thinking is something that most people are not well trained in. However, systems thinking is highly relevant not only in the current climate debate but also in the debates about regulation of financial markets and its implications for society. In this respect, science is expected to support and to become involved in processes of social learning in order to comply with these new demands (Siebenhüner 2004).

Especially in the climate change debate, systems thinking has gained new prominence within scientific discourse. In one way or another, we are forced to deal with complexities, with 'wholes' and with 'systems' in all fields of knowledge including energy policy, lifestyle changes or even in assigning the risks of genetic modification of organisms (e.g., to enhance carbon storage capacity). However, an issue so far not resolved is the question of how a society as a whole can learn systems thinking. This was also reflected in the discourse on this first question in the World System Café.

WHAT ARE THE ARENAS WHERE PEOPLE CAN LEARN SYSTEMS THINKING AND PUT KNOWLEDGE INTO ACTION?

The following statement summarises the key points of discussion on where to put the gained insights and knowledge into action. We present the main issues extracted from the discourse on this question by the World Systems Café participants.

Summary of Group Discourse

Statement 4:
There is no single arena, but rather several arenas that serve different purposes and stakeholders.

The groups identified several arenas that are relevant for learning systems thinking (Figure 2.2). These can be found on different scales (from individual and local to large scale and global) where different sets of issues can be tackled. Arenas include public print and audiovisual media, but the Internet can also provide a platform for e-learning. The success of science magazines or TV documentaries shows that there is a public interest in the world of science. This potential can be used to promote systems thinking. A good example for how cooperation between the media and the academic world can work in practice is the platform jointly created by the BBC and the Open University in the UK, providing a learning platform for "systems practice" (http://www.open2.net/systems/).

However, the groups pointed out that even a computer game (e.g., Nintendo, Xbox or Wii) has the potential to be a vehicle of learning. Apart from education or edutainment games, there are several commercial games like "Siedler" ("Settlers")[1], which also implicitly include aspects of systems thinking.

Other arenas where people can actively engage in putting knowledge on systems thinking into action are NGOs or local communities (e.g., Agenda 21 groups). Systemic issues such as climate change may have broad impacts on issues that environmental NGOs are concerned with. For instance, for a neighbourhood group, the causal loops that link local construction activities, such as the construction of a new shopping centre, car traffic and air quality, can draw attention to systems thinking and to the broader global

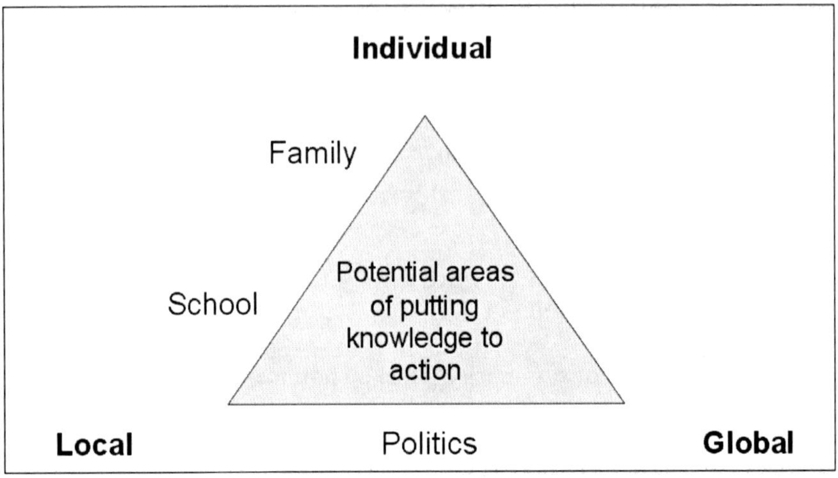

Figure 2.2 Arenas for learning systems thinking.

Discussion

The main motivation to broaden the scientific community and to include various forms of knowledge is based on the insight that many environmental and social problems, such as climate change—where actors and natural processes at the local, regional, national and global level interact—influence the causation of problems at hand (Siebenhüner 2004). In a world of ever-scarcer resources, a key challenge for scientists is thus how to increase the likelihood that our research will benefit wider society and the environment. This requires the development of more powerful research methodologies, generating knowledge from research and using this knowledge to inform action. Much of this is related to social learning. Thus, knowing is an act of participation in complex "social learning systems" (Wenger 2002). Beginning in the 1970s, the notion of social learning gained attention in many disciplines such as political science, in which the role played by advocacy coalitions in processes of societal change and learning was underlined (Tàbara and Pahl-Wostl 2007). Today, social learning can be regarded as an essential element of policy development and implementation. The sustainability science approach is especially dependent on the inclusion of participatory elements in knowledge production (Siebenhüner 2004). In participatory knowledge creation, systems thinking can provide a concept for linking scientific insights with local knowledge or stakeholders' insights.

In our view, social learning is tightly linked to systems thinking. Following Keen et al. (2005), social learning is defined here as the collective action and reflection that occurs among different individuals and groups as they work to improve the management of human and environmental interrelations. This must ultimately include all of us, since we are all part of the same system. Thus, knowing, learning and sharing knowledge are not abstract things we do for their own sake, but are part of belonging (Wenger 2002; Pahl-Wostl et al. 2008). However, how is public participation in knowledge production for systems thinking undertaken in practical terms? In all participatory arenas, the crucial question remains, 'Who participates?' The selection of participants can be viewed as a constitutional decision for the legitimacy, the course and the topics of each participatory procedure and its outcomes. Furthermore, the steps and approaches involved in this process bear high relevance for the success of the procedure, for example, informative methods versus co-ordination of different forms and fields of knowledge (Siebenhüner 2004). Therefore the generation and effective transmission of knowledge within societal systems is an overriding concern for the future. In this respect, a framework for understanding and promoting social learning systems for systems thinking must conceptualise learning as a social process. But what is learning from a social perspective? Wenger (2002) states that in a social learning system, competence is historically and socially defined and changes in a contradictory interplay with our own experiences. Through this dynamic two-way interplay learning takes place and action can emerge.

However, given that systems thinking feeds into sustainability, the question has to be raised to which extent participatory social learning procedures produce practical and sizeable outcomes and how far these lead to further social learning processes. What types of arenas are required to produce new knowledge or concrete products that are relevant to other societal groups and actors?

With respect to the analyses of the pressing environmental and social issues of our time, these should be accessible not only to well-trained decision-makers and academics but also to a majority of citizens. This implies a need for pragmatizing critical systems thinking so that it can be owned by citizens. Thus, citizens themselves ought to be able to apply basic critical systems ideas on their own behalf (Ulrich 1996). Ulrich argues that systems thinking has a potential to give new meaning to the concept of citizenship; it might enable all of us to become more responsible citizens. This is timely in the light of the future challenges to sustainability.

What are the arenas where systems thinking can be applied? Systems thinking can be relevant at the personal level when thinking for example about dietary habits, health issues or the way one acts in the working team. Systems thinking is considered as one of the central pillars of leaning organisations according to Senge (Senge and Sterman 1992). He sees it as one of the central disciplines besides personal mastery and team learning. These disciplines have found their way to change management and organisational transformation processes. Thus work can be an important arena for reflection on systems thinking.

Further key arenas include schools, universities and adult education. Besides these, entertainment and leisure activities carry potential for learning systems thinking too. Games, such as the climate board game "Keep Cool", foster systems thinking in a subtle way. Participatory exercises can also provide arenas for learning. Involvement of citizens in public consultation and planning processes provides possibilities for learning on systems.

WHAT TOOLS, METHODS AND APPROACHES CAN SCIENTISTS PROVIDE TO BRING SYSTEMS THINKING INTO SOCIETY?

This question was specifically raised in view of the problem of conveying learning for systems thinking in more practical ways into wider society. Within this broad scope, special emphasis was placed on tools such as modelling as one arena where system thinking can be learned. However, we explicitly do not mean to imply that social learning for systems thinking can only be conveyed by mathematical, deterministic approaches. On the contrary, it is evident that modern societies have moved away from traditional more holistic views. A return to a more encompassing "soft systems" thinking is timely.

Summary of Group Discourse

Statement 5: Tools = Models?

There was a general consensus that systems thinking can be regarded as a method that is applicable at multiple levels. At each level, a selection of different tools is required to investigate the system components that are chosen for the analysis. There was a strong consensus within the groups that computer modelling is a major tool supporting systems thinking. However, some members stressed that there are many other tools available apart from computer models and modelling. One weakness of addressing systems thinking by computer modelling is that the experience remains quite abstract. Through the discussion, the problem of the current dominance of linear thinking was raised several times. It was noted that systems thinking (in the sense of complex networks and cause–effect relationships) was not taught in schools. Intelligent computer games were suggested as one tool to overcome this gap. From the World Systems Café participants' responses we compiled a stakeholder tool matrix (Table 2.1). Herein the term 'tool' must be viewed in its broadest sense, that is, also as facilitating cultural belief systems as tools towards systems thinking.

Several members expressed concern over the current trend towards computer models. In their view, this holds the danger of totalitarianism by an overly strong faith in such models. Some models were perceived as

Table 2.1 Matrix of Stakeholder-Specific Tools for Systems Thinking

Tool	Application	Use of Computers	Suitable for	Example
Modelling	Simulating complex dynamic systems	YES	Schools, Academia	Vensim, Stella
Simulations	Visualizing complex dynamic systems	YES	Schools, Academia, Companies	Climate Sims
Role plays	Linking systems thinking with social interaction in a "safe" environment	YES/NO Has the potential to be computer-aided	Schools, rural communities, decision-makers, expert groups, etc.	Agriculture, Fishman, Fishbanks, SAMBA
Cultural belief systems	Steering and linking of system uses	NO	Individuals, communities	Taboos in natural resource uses
Agenda 21	Community-based discussion forum	NO	Communities	Regional/local planning initiatives
"Nintendo" computer games	Simulation games on linear system reactions	YES	Children/Adults	Die Siedler (Ubisoft)
GIS	visualisation of social-ecological systems	YES	Academia, Communities, State administration, regional planning, companies, etc.	Land-use planning, regional statistics, etc.

being strongly deterministic. It was argued that there was little room for manoeuvre left to develop own ideas of systems thinking if one employs predefined models that impose a given set of logic. There was, nevertheless, a strong consensus among participants that modelling provides different views on systems and thus has a positive effect on social learning on systems thinking.

Discussion

Coping with complexity is at the heart of the systems thinking approach. Several methodologies, such as, for example, agent-based modelling (Glaser, this volume) or quantitative models (Tretter and Halliday, this volume) have proven to possess extraordinary power in dealing with complex issues. Systems thinking is a general-purpose approach for modelling and simulation employed in order to deal with dynamic complexity. Thus, systems thinking and, specifically, research into system dynamics are grounded in control theory and the modern concepts of non-linear dynamics. For instance, cybernetics has brought forth structural models to enable organisations to deal with the complexity of their environments (Schwaninger 2004). In this sense, systems thinking is a paradigm of a specific worldview that shapes every aspect of how we experience the world. The systems theories and models we develop have their foundations in a number of elegant mathematical procedures (Sterman 2002). That being said, models are often called models of learning in the sense that they explain the emergence of equilibrium over time. To date, no widely accepted models of individual decision-making exist that provide the micro-foundations for empirical regularities. What are the motivations and learning rules used by citizens (Janssen and Ahn 2006)? Models try to endorse reality as a system; however, the reality is not a system but rather too complex and with emergent properties that are difficult to model. Thus, most learning models have been applied to rather simple games, which is understandable given that the formulation and testing of those learning models is often very complicated (Janssen and Ahn 2006). This results in tensions between qualitative systems thinking (e.g., rule-based modelling) and quantitative formal modelling (i.e., equation-based modelling). However, an increasing number of scholars are starting to confront their models with empirical observation in more rigorous ways (Janssen and Ostrom 2006).

For instance, Gurung et al. (2006) observed that the relationship between simulation models and collective decision-making in natural resource management has occupied a large part of the literature. However, the model is often a predominantly biophysical one, and few papers mention the participation of stakeholders from different organisational levels in the modelling steps. More recently, methods within modelling, such as "companion modelling", have demonstrated the ability of many illiterate people to use high-tech tools (Gonzalez 2000). In brief, the main principle of the companion modelling (ComMod) approach is to develop simulation models integrating various stakeholders' points of view and to use them within the context of the stakeholders' platform for collective learning (Röling 1996). This is a modelling approach in which stakeholders participate fully in the construction of models to improve their relevance and increase their use for the collective assessment of scenarios. The general objective of ComMod is to facilitate dialogue, shared learning and collective decision-making through interdisciplinary and "implicated" research to strengthen the adaptive management capacity of local communities (Gurung et al. 2006). Thus, such an approach would be of high interest to foster systems thinking within social learning for sustainability. Also in economic analysis, systems modelling approaches have been applied

(Kupers and Mangalagiu 2010). In contrast to general equilibrium models widely used for economic modelling, Kupers and Mangalagiu argue that it is possible to design models that mimic a set of human behaviours more closely to resemble those that occur in reality. To fully capture the dynamics of economics, the authors recommend including multi-scale effects by using agent-based modelling in dynamic models.

Difficulties remain regarding how to transfer knowledge on systems thinking into visual models of a system. For instance, geographic maps—one kind of static rigorous learning model of a system—are able to display only parts of a system, not the entirety. Borges (1998) has elaborated nicely on this tension in one of his short novels, "On Exactitude in Science":

> In that Empire, the Art of Cartography attained such Perfection that the map of a single Province occupied the entirety of a City, and the map of the Empire, the entirety of a Province. In time, those Unconscionable Maps no longer satisfied, and the Cartographers Guilds struck a Map of the Empire whose size was that of the Empire, and which coincided point for point with it. The following Generations, who were not so fond of the Study of Cartography as their Forebears had been, saw that that vast Map was Useless, and not without some Pitilessness was it, that they delivered it up to the Inclemencies of Sun and Winters. In the Deserts of the West, still today, there are Tattered Ruins of that Map, inhabited by Animals and Beggars; in all the Land there is no other Relic of the Disciplines of Geography . . . (Suárez Miranda, Viajes de varones prudentes, Libro IV, Cap. XLV, Lérida, 1658) (Borges 1998: 325)

In today's virtual worlds of Second Life, Facebook, Twitter, and so on we can see some parallels to this story of poetry. They reflect and construct a parallel virtual world and thus can be regarded as a model of the 'real world' in some ways. Albeit, whether such platforms or tools have the potential of increasing our understanding of complex interactions in the real world remains to be seen. It appears though, that it is necessary to conflate qualitative and quantitative modelling and design approaches to support relevant systems thinking knowledge production at different levels in society.

WHAT DO YOU NOT PERCEIVE OF THE WORLD WHEN YOU APPLY SYSTEMS THINKING?

This question aimed to address the problem of drawing conscious boundaries. This is at the heart of systems thinking methodology. All system boundaries are artificial and have certain sets of values attached to them. They need to be constructed to reflect the system of interest appropriately.

Summary of Group Discourse

Statement 6:
Values, motivations and emotions are left out of the focus when applying systems thinking.

Systems thinking on the individual level includes personal relations aspects that cannot be encapsulated in a whole-systems approach. However, systems thinking can be viewed as a stepping-stone for emotions. This is done intuitively, since the cognitive apparatus of the human brain is not capable of assimilating complex issues without training. That said, accreditation in the teaching curricula is timely to face the challenges of sustainability. Furthermore, systems thinking can be regarded as a tool to broaden the horizon—the more you know and the more you can connect seemingly unrelated issues, the larger is your horizon. A key concern that was raised at length in all groups was the issue of system boundaries: We do not see what is left out in our model of a system. These boundaries can be highly random and thus may not reflect the system of interest appropriately. Thus, creating artificial system boundaries may be necessary in order to be able to investigate and understand a particular sub-system. Since it is a reproduction of the world, it only covers a part of it. The crucial issue is to find the necessary level of detail (for the problem or issue at hand) and to define links to other sub-systems.

A further aspect that was highlighted was that spontaneity and irrationality is easily excluded when using the systems' lenses. Several participants of the World Systems Café exercise stressed that systems thinking does not recognise beauty and is most often connected with the notion of mechanising and formalising the world. Then, the discussion revolved around the question whether emotions and personal values can or cannot be quantified by systems thinking and how poetry fits into it. It was debated whether the Rational Actor Paradigm (RAP), which lies behind neoclassical economic theory and many economic modelling efforts, can be considered to represent human behaviour in an adequate way or if it is reductionistic and simplistic. Many scholars argue that this paradigm is simply wrong, since human beings are not rational actors. The question then emerged in this session, whether systems thinking in this sense might be too rational an approach.

This question was further discussed on two other levels: first, whether irrational behaviour of a system or actor can be depicted by systems analysis, and second, whether in sustainability discourse, rational argumentations are effective at all. Irrational behaviour can be depicted by systems thinking, for example, in simulations such as the Beer Game (a widely applied management simulation exercise). If we enter the field of preferences (e.g., buying cigarettes) and reflexive preferences (e.g., wanting to stop smoking), the power of systems thinking rather lies in approaches such as systems therapy (Tretter and Halliday, this volume). Regarding the second-level question, it was noted that successful sustainability communication requires emotions, images and people to relate to. As a case in point, one group gave the example of the movie "An Inconvenient Truth" as a success. They argued that the movie reflects this issue quite well, since the charisma of Al Gore and the visualisation of climate impacts and other science contents were the key to the wide reception of this systems thinking topic.

Two major stumbling stones were identified in this line of argument. First, that if there is the feeling of understanding the system, the notion may emerge that one is able to control the system. Second, by mechanising, formalising and quantifying the world in a systems' framework, the risk arises that issues will be addressed from too rational a point of view. The discussion closed with the recognition that through the application of systems thinking it is likely to see more possibilities, therefore more choices may emerge.

Discussion

There is a very long tradition that places values at the heart of systems thinking—for example, Lazlo (1972), "[T]here is nothing in all the realms of natural systems which would be value free when looked at from the vantage point of the systems themselves." In contrast to this, the participants felt that values are not part of a systems thinking approach and key authors, such as Churchman (1982) and Checkland (1984), saw "moral outrage" as the foundation stone of systems thinking. This discrepancy might be explained by the strong focus of the former discussion rounds at the World Systems Café on methods and models that have a strong rational origin. However, the danger of looking only at the technical or the quantitative side of systems thinking is that it is inconsistent with the spirit and goals of systems thinking, by underestimating people's interests and capabilities. Systems thinking requires respect and empathy for others and for other viewpoints (Sterman 2002). Thus, the problem of social learning for sustainability lies, in part, with failing to recognise the importance of prevailing mental models and values. These affect the view on systems thinking. Thus, the more profound the change in strategy towards sustainability, the deeper must be the change in thinking. Indeed, it can be argued that improving and changing the mental models in society is the fundamental task of systems thinking for sustainability. However, challenging mental models is delicate, especially if inconsistencies between assumptions about the structure of the system and its likely behaviour are revealed (Senge and Sterman, 1992). If trust and openness are not well established, individuals may be threatened and react defensively. Thus learning should be a mutual process in which the concerns and the starting point of the journey to systems dynamics are taken into account.

In this context, the movie "An Inconvenient Truth" by Al Gore is an interesting case. Being a documentary on climate change, it contains a great deal of science talk as well as reflections on how (climate) systems behave.

Figure 2.3 Vulnerability (Copyright © 2003 Hannah Förster. www.minimalista.org. hannah.f@gmail.com).

Why has this movie become so successful worldwide and contributed to awareness building, particularly in the US and in Asia? The success factor was presumably not systems thinking but rather the thematic problem approach, combined with the charisma of the presenter, Al Gore. Thus, systems thinking for sustainability seems to require an emotional driver to reach people.

A piece of art such as an image (Figure 2.3) may convey an emotionally stronger representation of the fragility of the earth system, and may be inspired by systems thinking.

SYNTHESIS

What lessons can be extracted from the World Systems Café exercise at Sommerhausen? In essence, the World Systems Café discussions uncovered the hidden assumptions and biases each of us has in terms of systems thinking. Thus, if we recapitulate our initial views when formulating the title of the World Systems Café and the respective questions, we feel that the title of this chapter is partly misleading. The discussions within the World Systems Café exercise indicated that the background, experiences and personal knowledge of each individual seem to determine the definitions and views on systems thinking.

Yet tools may help to streamline some of the diverging initial definitions and ideas expressed at the conference. The tool matrix (Table 2.1) gives an exemplary view on the kind of tools that can be applied to teach and train people in systems thinking. Simple bathtub simulations, such as the Climate Sims (http://sustainer.org/tools_resources/climatebathtubsim.html) developed by the Sustainability Institute in collaboration with MIT, Schlumberger and others, may convey the central messages. Initial experiences with distance learning are encouraging and the feedback from students showed that systems thinking can be acquired with this kind of interactive simulations.

The human societies of today developed into multifaceted highly fragmented and diverse entities with a broad array of different interests, views, knowledge structures, perspectives, norms and values (Becker and Jahn 1999). This makes social learning for systems thinking so difficult to tackle.

In the past, sustainability issues and social learning about these were seldom the focus of interest, as the era of the Industrial Revolution removed people far from nature (Herod 2009). However, a perception transition has taken place ever since the reports of the Club of Rome in 1972, Brundtland Report 1987, the 1992 Earth Summit in Rio and its follow-up in 2002 in Johannesburg. Thus, awareness of the disconnection of man and nature is increasing, for example, due to high energy prices and carbon constraints, which force companies to think about ways to save energy and to reduce their intensity of resource use. Reflecting on the impacts of companies' actions in a Future Search Conference (Weisbord 1993), or in a Real Time Strategic Change exercise (Jacobs 1994), may lead to new visions, which question traditional ways of doing things. Communicating convincingly that a product or service is produced sustainable is a challenge for today's public relations departments. To take into account all the effects of our actions on other parts of the world or in the distant future (e.g., toxic materials that may accumulate in humans and animals) requires systems thinking and systems communication (Renesch 2008). Thus, it seems imperative to create the conditions for learning experiences that re-establish the mental connections between people's actions and environments via systems thinking and social learning.

The systems thinking community is very diverse. Different scientific communities have developed their own vocabulary, observation methods, concepts and models, some of which have been adopted by other communities and have become part of systems thinking in various ways.

It would be worthwhile to bring these heterogeneous communities in closer contact with the communities engaged in social learning, organisational learning and pedagogics. An example from the field of organisational learning is the Society for Organisational Learning (SoL), which places its emphasis on systems thinking, considering it to be the "fifth" discipline alongside "personal mastery", "mental models", "building shared vision" and "team learning". Of these, the key discipline is systems thinking, which helps to overcome learning barriers. Furthermore, it fosters the ability to see structures and patterns rather than singular events. The activities of the SoL focus not only on companies but on any kind of organisation including schools.

The discussion that surfaced in the World Systems Café on the topic of systems thinking for sustainability point to the unresolved complexity between natural, social, economic and cultural factors. This can be attributed to the fact that historically, systems thinking was driven mainly by two traditions (originating from the social and natural strands) that both had limited scientific aspirations in conceptual, theoretical and methodological terms to link them towards sustainability.

Since the challenge of integrating natural and social science research in a comprehensive manner in systems thinking remains, lessons learned from the Sommerhausen experience may be converted into criteria for the elaboration of a systems thinking framework for sustainability.

CONCLUSIONS

The fact that interdisciplinary teams are more common in natural resource management research centres now reflects the current understanding that no natural environment is exempt from human inputs. Thus, we must examine the system as a whole, including the human component, from a systems thinking perspective if we wish to protect and manage it in a sustainable way.

The emphasis hence has moved from the need to simply 'know more' and deploying even more information to policy and expert circles towards the development of adaptive cross-sector capacities and new types of knowledge. Such knowledge in our view allows a more adequate response to the changing dynamics of social–ecological systems in concrete contexts of action.

However, as surfaced in the discussions in the World Systems Café in Sommerhausen, the difficulties of systems thinking as part of a process of social learning for sustainability lie not so much in our lack of understanding of the functioning of ecological systems, but in our lack of understanding of the governance of socioeconomic and cultural systems. Key questions include how these are structured and managed and how they interact with ecological systems.

New institutional approaches are required to promote systems thinking as part of social learning for sustainability. These must be based on a new framing of the issues at stake and the inclusion of stakeholders. These range from school children, students and adults to the policy-makers in the political and economic arenas.

It was a consensus at the World Systems Café that learning systems thinking should start at school level and continue throughout an adult's working life including further education. The example about mitigating climate change which was used in the beginning of this chapter is a classic example where systems thinking is required.

Without a public understanding of the necessity to act quickly in order to avoid something much more severe in the future, radical reductions of

CO_2 emissions cannot be achieved by the international community in the near future. Some basic understanding of systems thinking embedded in the society is thus mandatory.

Systems thinking is, on the one hand, needed to be able to understand how the global climate system with all its subsystems works. Systems thinking is however equally important when searching for the response strategies to combat climate change, poverty and many other pressing global change problems. For finding the right leverage points in society, in the economic system, ecological systems and in politics, the systems thinking approach thus provides helpful tools and a paradigm for a transition towards sustainability.

ACKNOWLEDGMENTS

The participants from the World Systems Café are thanked for their willingness to engage so actively after the long morning sessions of the conference during a very hot and humid afternoon. The parish hall is thanked for allowing us the use of their piano so that we could provide the participants with a 'real' Café atmosphere. Andrew Halliday is thanked for his helpful comments and his willingness to share his in-depth knowledge on the different schools of thought on systems thinking. We thank Hannah Förster for providing the nice piece of art image on the fragility of the earth system. Last but not least, Egon Becker, Felix Tretter and Robert M. Griffin are thanked for their critical revision of the manuscript.

NOTES

1. In this particular game the player creates new settlements and has an impact on the surrounding landscapes. For instance, cutting too much timber for construction purposes leads to deforestation and thus to environmental degradation, which, in turn has an impact on the food supply of the settlements.

REFERENCES

Arias Schreiber, M, Glaser, M (2008) Systems thinking: Old ideas for interdisciplinarity in marine social-ecological systems. Proceedings of the FAO Conference on "Coping with Global Change in Marine Social-Ecological systems". FAO, Rome

Becker, E, Jahn, T (1999) *Sustainability and the Social Sciences: A Cross-Disciplinary Approach to Integrating Environmental Considerations into Theoretical Reorientation.* Zed Books, London, New York

Borges, JL (1998) On exactitude in science. In Borges, JL (ed) *Collected Fictions.* Penguin Books, London

Brown, J, Isaacs, D (2005) *The World Café: Shaping Our Futures Through Conversations That Matter.* Berrett-Koehler Publishers Inc., San Francisco

Caulkin, S (2006) Why things fell apart for joined-up thinking. *The Observer,* Sunday February 26, 2006. Guardian News and Media Limited, London

Checkland, P (1984) *Systems Thinking, Systems Practice.* Wiley & Sons, Chichester, New York

Churchman, CW (1982) *Thought and Wisdom.* Intersystems Publications, Seaside

Gonzalez, R (2000) Platforms and terraces: Bridging participation and GIS in joint learning for watershed management with the Ifugaos of the Philippines. Dissertation, ITC-Wageningen University, Wageningen

Gurung, TR, Bousquet, F, Trébuil, G (2006) Companion modeling, conflict resolution, and institution building: Sharing irrigation water in the Lingmuteychu Watershed, Bhutan. Ecology and Society 11: Art. 36 (online)

Hall, AD, Fagan, RE (1956) Definition of system. Yearbook of the Society for the Advancement of General System Theory 7: 18–28

Herod, A (2009) *Geographies of Globalization.* Wiley-Blackwell, Oxford

Jacobs, R (1994) *Real Time Strategic Change: How to Involve an Entire Organization in Fast and Far-Reaching Change.* Berrett-Koehler Publishers Inc., San Francisco

Janssen, MA, Ahn, TK (2006) Learning, signaling, and social preferences in public-good games. Ecology and Society 11: Art. 21 (online)

Janssen, MA, Ostrom, E (2006) Empirically based, agent-based models. Ecology and Society 11: Art. 37 (online)

Kates, RW Clark, WC, Corell, R, Hall, JM, Jaeger, CC, Lowe, I, McCarthy, JJ, Schellnhuber, HJ, Bolin, B, Dickson, NM, Faucheux, S, Gallopin, GC, Grübler, A, Huntley, B, Jäger, J, Jodha, NS, Kasperson, RE, Mabogunje, A, Matson, P, Mooney, H, Moore III, B, O'Riordan, T, Svedin, U (2001) Sustainability science. Science 27: 641–642

Keen, M, Brown, V, Dyball, R (2005) *Social Learning in Environmental Management: Towards a Sustainable Future.* Earthscan, London

Kupers, R, Mangalagiu, D (2010) Climate change policy: Positive or negative economic impact? Why? European Climate Forum (ECF) Working Paper 1 (online)

Laszlo, E (1972) *The Systems View of the World: The Natural Philosophy of the New Developments in the Sciences.* Braziller, New York

Midgley, G (2003) *Systems Thinking.* SAGE Publications, London, Thousand Oaks, New Delhi

Pahl-Wostl, C, Mostert, E, Tàbara, D (2008) The growing importance of social learning in water resources management and sustainability science. Ecology and Society 13: Art. 24 (online)

Ramanathan, V, Feng, Y (2008) On avoiding dangerous anthropogenic interference with the climate system: Formidable challenges ahead. Proceedings of the National Academy of Sciences 105: 14245–14250

Renesch, J (2008) A systems view of corporate culture change. 11 June, first appeared in February (online).

Röling, N (1999) Modelling the soft side of the land: the potential of multi-agent systems. In: Leeuwis, C (ed) *Integral Design: Innovation in Agriculture and Resource Management.* Mansholt Institute, Wageningen, pp 73–97

Schwaninger, M (2004) Methodologies in conflict: Achieving synergies between system dynamics and organizational cybernetics. Systems Research and Behavioral Science 21: 411–431

Senge, PM, Sterman, JD (1992) Systems thinking and organizational learning: Acting locally and thinking globally in the organization of the future. European Journal of Operational Research 59: 137–150

Siebenhüner, B (2004) Social learning and sustainability science: Which role can stakeholder participation play? International Journal of Sustainable Development 7: 146–163

Sterman, JD (1994) Learning in and about complex systems. System Dynamics Review 10: 291–330

Sterman, JD (2002) All models are wrong: Reflections on becoming a systems scientist (Jay Wright Forrester Prize Lecture). System Dynamics Review 18: 501–531

Sterman, JD (2008) Risk communication on climate: Mental models and mass balance. Science 322: 532–533

Tàbara, JD, Pahl-Wostl, C (2007) Sustainability learning in natural resource use and management. Ecology and Society 12: Art. 3 (online)

Ulrich, W (1996) Critical systems thinking for citizens. In: Flood, RL, Romm, NRA (eds) *Critical Systems Thinking*. Springer Verlag, Heidelberg, pp 165–178

von Bertalanffy, L (1968) *General Systems Theory: Foundations, Development, Applications*. George Braziller, New York

Weisbord, MR (1993) *Discovering Common Ground: How Future Search Conferences Bring People Together to Achieve Breakthrough Innovation, Empowerment, Shared Vision and Collaborative Action*. Berret-Koehler Publishers, San Francisco

Wenger, E (2002) Communities of practice and social learning systems. Organization 7: 225–246

Part II

Social-Ecological Systems (SES) and Complexity

3 Social-Ecological Systems as Epistemic Objects

Egon Becker

INTRODUCTION

During the last decade, the concept of social-ecological systems (SES) has become central to an increasingly widespread international discourse on human–nature interactions. In this discourse, the Resilience Alliance, a Stockholm-based research association that includes several important institutes, has played a leading role, in defining both the concepts and the aims of research on SES. This research focuses mainly on the adaptive management of ecosystems from the perspective of resilience. Resilience is understood here as the capability of a system to retain similar structures and functioning after disturbances, thus ensuring continuous development (Folke 2006).

Still, the SES concept has been used in other research areas as well, ones with different theoretical backgrounds, and different aims and objects. All of these share the conviction, however, that human activities have a strong and formative impact on the earth's ecosystems, climate and hydrosphere—a claim encapsulated in the notion that we have entered a new geological epoch, the so-called Anthropocene. The various participants in the discourse on human–nature interactions have, moreover, all tried to translate their general convictions into goal-oriented research activities. We find here a broad spectrum of authors employing a wide variety of related concepts.

Recently, Brand and Jax (2007) have taken a critical and sceptical look at the redefinition and extension of the concept of resilience, which has been stretched from a descriptive-analytical term in biological ecology into a vaguer and more malleable notion, one supposedly able to function as a general approach to systems analysis that can be used by different scientific disciplines and research fields. In this latter sense the concept of resilience is being used, on the one hand, pragmatically as a communication tool linking different scientific disciplines and, at the same time, to span the gap between science and practice. On the other hand, the concept is being used as a point of reference for a redefinition within the cognitive context of individual scientific communities. Thus, referring to SES, Brand and Jax note that resilience has been increasingly viewed as a general perspective, as a way of thinking about complex systems. But this extended use of resilience provides no clear definition of the concept (Anderies et al. 2006). In other words, resilience has been transformed from an epistemic object with a well-defined meaning and use into a weakly structured boundary object situated at the intersection of a large number of disciplines and research fields.

A similar analysis can be made of the concept of SES, still the most important object of resilience research. We can understand this concept from three different perspectives:

1. SES as boundary objects situated at the intersections of individual fields of research and disciplinary settings
2. SES as epistemic objects, as 'things' that humans can and want to know about using well-defined methods of research and theoretical reasoning
3. SES as real objects represented in system models constructed for dealing with problems and phenomena in various fields of application

In the following, the SES concept will be examined from an epistemological point of view that is marked by a critical theory of societal relations to nature[1]. Critical here indicates that it is necessary to take a hard look at the consequences of making conceptual distinctions and of holding ontological convictions. For example, in order to have an adequate understanding of SES a constitutive distinction must be made between nature and society[2]. The term 'societal relations to nature', on the other hand, refers to a need to adjust one's views concerning the patterns of connections between analytically distinguished entities.

ANTHROPOCENE: THE END OF PRISTINE NATURE AND A DENATURALISED SOCIETY

Human activities alter and shape the planet earth, entangling local environments in global bio-geochemical cycles and geophysical processes. Many of the earth's ecosystems are dominated directly by humanity, and no local or regional ecosystem on earth's surface is free of pervasive human influence. However, the rates, scales, kinds and combinations of changes today are fundamentally different from those at any other time in history. "We are changing Earth more rapidly than we are understanding it" (Vitousek et al. 1997).

Reflecting on the ever-increasing human impact on the earth's bio-, geo-, hydro- and atmospheres, the Nobel Prize–winning chemist Paul Crutzen (2002) has suggested we have entered a new geological epoch, which he calls the Anthropocene. Crutzen's term has received increasing acceptance, including that of influential geologists (Zalasiewicz et al. 2008). At the same time, both human societies and globally interconnected economies increasingly depend on ecosystem services[3] and the maintenance of ecosystem functions. This manifold of systemic interdependencies among natural and social processes, occurring at different temporal and spatial scales, demands an appropriate conceptual frame.

A New Epistemic Constellation

If we take the idea of an Anthropocene seriously, then we must recognise that a new epistemic constellation has emerged, one marked by complex human–nature relations at its centre (Liu et al. 2007). Such complex relations have played a critical role in numerous studies and reports on global change and climate research (IPCC 2007), in earth-system analyses (Schellnhuber et al. 2004) and in sustainability science (Kates et al. 2001). The conclusion seems unavoidable: In the Anthropocene, it is impossible to understand nature without society, and society without nature, as the German sociologist Ulrich Beck (1986, 1992) pointed out persuasively more than 20 years ago. Moreover, if we accept this point, then both the idea of a pristine nature and a denaturalised society are outdated, and the academic separation of science from the humanities, and, within science, the social sciences from the natural, proves to be a serious obstacle to the progress of science and to the relevance of science for solving urgent social-ecological problems.

A New Description of the World

In the last two decades, concerned scientists in quite different fields of research have come to agree more and more on a few general principles to govern a new description of the world we are living in:

- The planet earth as a whole is a crisis ridden self-organising complex system.
- Mankind is an integrated part and a powerful driver of the earth's systems dynamics.

While this seems straightforward enough, two consequences of this emerging new worldview are less obvious, and science has only very rarely seemed aware of the epistemic abyss these ideas open up:

If mankind is an integrated part of the complex system earth so is the scientific observer, which means that observations are only possible from inside the system, made by a participant observer.

If human action is guided by divergent values, and by everyday and scientific descriptions of problems and phenomena, these are also part of the observed reality: The system is self-describing and self-referential.

With a few exceptions such as Niklas Luhmann's (1995) theory of autopoietic social systems, the theoretical implications of recognizing that an observer is a part of the system observed, a participant observer, and that system processes are necessarily self-referential, has rarely been an issue of debate in mainstream science. Yet if general principles of the new worldview are also to be available to the environmental sciences and to human ecology, theory building there requires a transfer of ideas, concepts, methods and models of systems from general systems theory, complexity theory, autopoiesis theory[4], actor-network theory and network topology.

These avant-garde fields of research and modes of thinking are disrupting the accepted discourse on human–nature relations. In these fields, science is moving from things to relations, from structures to processes, and from identity to difference. These shifts in basic categories of thinking and understanding have infiltrated the academic world, and also art and literature. Hans-Peter Dürr (2001) summarises the new worldview in the aphorism, "A stone is a coagulated pattern of dynamic relationships." For the theory of societal relations to nature they provide a general orientation for thinking and a guideline for concept formation (Becker and Jahn 2003, 2006).

SOCIETAL RELATIONS TO NATURE

The new description of the world mentioned previously, together with the centrality of human–nature interactions and the shift in general categories of theoretical thinking, call for a theoretical concept that comprises a mediation between natural and social entities. In the 19th century, Karl Marx argued in a radical fashion against any kind of reification of social relationships into fixed things: "Society does not consist of individuals but expresses the sum of interrelations, the relations within which these individuals stand" (Marx 1857: 176). In an analogous way we can describe the various interactions that take place between natural and social entities as constituting 'societal relations to nature'. Such relations must be regulated in every society in such a way so as to sustain the cross-generational continuation of societal processes necessary for life; without such regulation, these processes collapse. The regulation of material and energy flows, however, is linked to a multiplicity of cultural symbolizations and embedded thereby in societal structures and processes of communication.

Societal relations to nature, therefore, are symbolically mediated, material-energetic patterns of regulation. This general definition does not tell us, however, at which level the nexus of regulated relations is to be located, or where research is to be carried out. Here we can just say that societal relations to nature take form both directly through the interaction of individual human actions and at the level of institutions and differentiated functional systems. With this as a start let us now examine them more closely.

A Definition of Societal Relations to Nature

The concept of societal relations to nature stands in opposition to the mainstream tradition in European philosophical thinking that is centred on the category of substance. Instead it opts for an "ontology of relations and processes", to quote Alfred North Whitehead (1978). Societal relations to nature are the historically and culturally specific patterns and practices by means of which societies attempt to materially regulate, and culturally symbolise, their various relationships to nature. They emerge from a nexus of causal effects ("Wirkungsgefüge") embedded in a field of symbolic meaning ("Deutungszusammenhang"). Therefore they always exist as intertwined physical and symbolic forms.

Following the discussion of 'core basic needs' within the sustainability discourse we will call work and production, sexuality and reproduction, nutrition and water supply, mobility and housing basic societal relations to nature because they are relations that are indispensable for both individuals and societies. If these relations are not secured permanently the reproduction and development of societal life will fail.

Forms of Regulation

Taking a closer look at the forms and practices in which and through which humans and societies attempt to materially regulate and culturally symbolise their relationships to nature, 'societal relations to nature' can be further refined. Specific forms of the regulation of basic relationships take shape at different levels:

- Micro-level: regulation of the cultural forms of individual satisfaction of needs and provisioning activities
- Meso-level: regulation of supply systems, resource utilization in institutional contexts
- Macro-level: regulation of societal reproduction and social integration as congealed into production, property and gender relations

The concept of 'societal relations to nature' emphasises the difference between material regulations and cultural symbolizations, and it also underlines the importance of 'hybrid objects', objects that are both natural and cultural (Latour 2005).

A CONTESTED TERRAIN: WORLDVIEWS AND THE TRANSFORMATION OF BOUNDARY OBJECTS

If we look at the academic world and ask, "In which domain are studies on human–nature interactions carried out?" we obtain a long list of coexisting and competing disciplines, multi- and interdisciplinary research fields, human ecological areas and international programmes, all of which operate more or less in isolation within different scientific cultures and professional worlds, working parallel to one another with divergent points of view, with little exchange of concepts and methods, and with a confusing jumble of terminology. Anyone peeking over the fence of his or her own discipline or research field soon comes to realise that no stable convention exists with regard to the meaning or definition of the term 'human–nature interactions'.

Given this situation, it won't do to advocate a royal road of building a general theory of everything; rather, what is needed are concrete proposals for building, linking and activating scholarly networks. Using financial incentives and the evaluation of projects, science policy and research funding agencies have tried again and again to bring together researchers from different fields in problem and goal-oriented research associations. Although burdened with a widespread feeling that the situation is still unsatisfactory, these practical activities nevertheless direct scientific attention to common topics and interests in miscellaneous fields of research. But there remains, for the most part, no consensus over concepts, methods and objects. Yet the extent to which and the manner in which different professional groups collaborate in scholarly networks, and how research results arising from often radically different worldviews and theoretical orientations are incorporated into a coherent ensemble, remains a critical issue; without such translation and integration of research results, information from one study may not be available for or transferable to other studies.

Social and Cognitive Function of Boundary Objects

The conditions governing success in multi- and interdisciplinary research have been investigated in numerous case studies within the history and sociology of science. These studies have produced several analytic tools that have enabled a better understanding of the transfer of concepts between heterogeneous scientific groups and discourses, as well as cooperation among these groups. For example, Susan Leigh Star and James R. Griesemer (1989) have shown that standardisation of methods and the development of boundary objects are major factors in efficient cooperation among heterogeneous scientific communities and between the latter and non-scientific actors.

Standardization of research methods is directed against an unprincipled eclecticism. It serves to 'discipline' the information and results coming from different researchers, making them compatible with and translatable into different scientific communities. Boundary objects, on the other hand, are objects, either concrete or conceptual, that are situated and developed in the borderlands between heterogeneous discourses. Given their divergent aims, interests and practical and theoretical orientations, boundary objects are defined only loosely and generally. Moreover, they are multifunctional. They have a social function as a tool for communication and cooperation among different scientific communities, who can all agree that they are talking about the same thing, and yet still attach different meaning to this thing. At the same time, boundary objects also have a cognitive function, acting as 'trading zones' for a transfer of concepts and methods (Galison 1991).

According to Star and Griesemer (1989: 393) boundary objects are "both plastic enough to adapt to the local needs and constraints of the several parties employing them, yet robust enough to maintain a common identity across sites. They are weakly structured in common use, but become strongly structured when used at individual sites." They may be abstract or concrete and, although they may have different meanings in different social worlds, their structure is common enough to more than one world to make them recognizable across borders, thus functioning as a means of translation. "The creation and management of boundary objects is a key process in developing and maintaining coherence across intersecting social worlds."

Social-Ecological Systems as Boundary Objects

The question now is whether there is a common object (concrete or conceptual) to be found within the border zones connecting the wide areas of research on environmental issues, one that can serve research activities within individual disciplinary areas as well. In human ecology, the concept human–nature interactions suggests itself as an appealing candidate. In the Frankfurt version of social ecology, this concept has been reworked and refined in terms of societal relations to nature (Becker and Jahn 2006), while in the Vienna version it has been developed into a concept of societal metabolism (Fischer-Kowalski and Weisz 1999), with other conceptual reformulations appearing elsewhere. But as a possible boundary object human–nature interaction is obviously too general and too abstract. A daunting epistemic task faces anyone wishing to conceptualise 'interaction' and 'relation' as objects, and it is difficult to imagine how an effective ordering and centralization of heterogeneous discourses would be possible in this way.

Global change and sustainable development might seem to be strong candidates as well. Both are used in science, politics and public debates, and both have a strong potential for consensus building. As concepts they refer to both qualities of processes and normative options. However, neither refers to a subject of change or development; therefore in research on global change the earth system takes on the role of a boundary object, and one studies how the earth system can and will develop within corridors of sustainability (Schellnhuber et al. 2004). But here the emphasis is on research at a global system level and thus the concept of the earth system is only of restricted use for research at local or regional levels.

The situation does indeed seem to improve if human–nature interactions are conceptualised in terms of systems, for 'system' seems to be more concrete than 'relations'. Systems, in fact, include relations, emphasising the patterns among them, thus referring also to their boundaries. Given these advantages, it is not surprising that one finds many—too many, in fact—good system concepts vying as candidates to play the role of a boundary object. In each instance, a constitutive distinction between nature and society (or nature and culture, etc.) is made and the system is endowed with specific properties (complex, self-organised, adaptive, etc.).

The results have so far not been encouraging. A disorderly diversity of names and concepts has been produced, and yet it is not clear whether the one name is being used to refer to the same thing, or whether the same thing is being given different names. In my view, the concept of SES has proven itself the strongest and most convincing candidate in the contest for a boundary object relevant both to sustainability science and to the study of the manifold of interdependencies among natural and social processes along different temporal and spatial scales. One powerful advantage of the SES concept is that it can assimilate the other strong candidates (e.g., earth system, world system, human–environment system, managed ecosystem), thus providing translation possibilities across many divergent discourses.

Human ecology, cultural ecology, social ecology, environmental science, sustainability research and research on human dimensions of global change should be able to reach a working agreement on SES as a common boundary object without major conflicts. Such an agreement would mark more than just a superficial terminological accord, since SES as a boundary object would possess a weak structure, thus making varied uses possible.

Transformation of Boundary Objects into Epistemic Objects

In the history and sociology of science, the idea of a boundary object has been used for the reconstruction of completed and successful research processes. However, research on SES is still in its infancy. Therefore it needs a boundary object that can function as a pragmatic tool, one capable of stimulating cooperation and communication among separate areas of research. In addition to this social function, SES as a boundary object must have a cognitive function for the research process as well. To achieve this cognitive function SES as a boundary object must be restructured into an epistemic object that is capable of being used at individual sites.

The concept of 'epistemic object' derives from the history of science (Rheinberger 1997, 2006). It refers to a 'thing' that humans can and want to know. Conceptual distinctions, experimental setups, fields of applications, methods of observation, mathematical models and forms of data presentation all constitute epistemic objects. The transformation of a boundary object into an epistemic object is guided, explicitly or implicitly, by preanalytic ideas, general world views and ontological convictions (often represented in the form of simple mind maps).

Within the broad discourse on human–nature interactions, SES can be characterised as boundary objects by several general properties:

- As systems, they consist of elements, the relations between them, and the borders delimiting the system.

- As social-ecological units, their elements (and their relations) are classified as either 'social' or 'natural' or 'hybrid'.
- Additionally, they get marked as complex systems—that is, they behave non-linearly; they have positive and negative feedback loops; they may form hierarchies, thus displaying emergence and self-organisation; and finally they depend strongly on their context and history (see Ratter, this volume).

If such a complex SES is transformed into an epistemic object, these general properties are reformulated as hypotheses about specific units of research.

SES as epistemic objects also embody the non-knowledge related to them. In this way, they are linked to problems concerning knowledge or action, and therefore they require methods and instruments for bridging the gap between knowledge that is available and knowledge that is needed. Knowledge, problems and methods are the constitutive components of an epistemic object. Research on SES involves the case-by-case transformation of these components (Becker 2006).

When a boundary object is strongly structured for individual site use its components become defined within the cognitive frame and the social conditions of a specific scientific or professional community. In Figure 3.1,

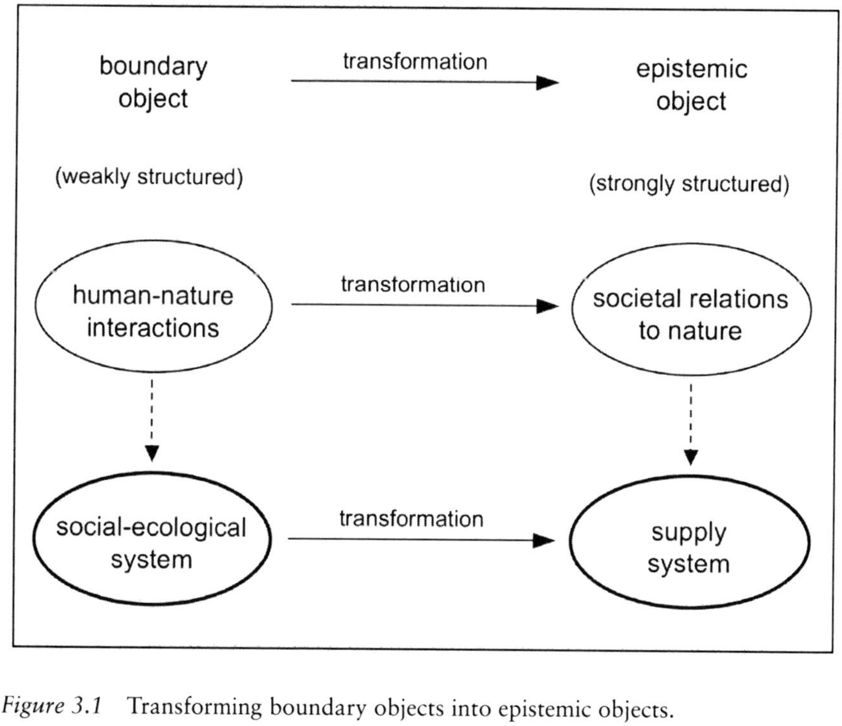

Figure 3.1 Transforming boundary objects into epistemic objects.

the vertical arrow on the left side symbolises the redefinition of the concept 'human–nature interaction' into the weakly structured general boundary object 'SES', while the horizontal arrows mark (in this example) the transformation of this boundary object into an epistemic object within the context of (Frankfurt) social ecology by means of a strong restructuring. Thus, first, the concept of 'human–nature interaction' is transformed into the concept 'societal relations to nature'; second, this latter concept then is exemplarily restricted to the concept 'supply system' formulated as an SES (Hummel, in this volume).

A WORLD OF COMPLEX SYSTEMS

It is a commonplace that complex systems are difficult to imagine and to describe. Describing complex systems centred on human–nature interactions is difficult therefore not only because the effort is relatively new; it is also an inherently challenging task. To begin with, a distinction must be made between nature and human society (or nature and culture). Without such a distinction, the interaction between them is unthinkable.

An analytical distinction between nature and culture, between a physical world and a mental world, has, of course, been fundamental to the European worldview since ancient times. Here one must be careful, however, not to turn an analytical distinction into an ontological separation, as did Descartes and his successors. Rather, the distinction marks a semantic difference; it is a prerequisite of rational thinking (Habermas 2005: 155ff). Though analytic, this distinction still represents a deep conceptual division. Attempts at reducing complexity following this initial cut come with a high price.

Following this distinction between nature and culture, physical world and mental world, SES are situated in a no-man's-land between two analytically separated domains and also between the academic cultures of natural and social sciences. Thus they have come to be classified not as either natural or social but rather as both natural and social. In the view of Bruno Latour (2005) they are always hybrids.

Formal Definition of a System

Simply re-naming the separated domains 'natural system' and 'social system' won't do because qualifying both as a 'system' carries a heavy logical burden. Even if we define (with Liu et al. 2007) SES as coupled human/social and natural systems, the meaning of the term 'system' still has to be explained; otherwise 'system' is just a metaphor for a compound of things. For the moment, however, we can simply note that construing SES as epistemic objects requires making a distinction between nature and society that refers either to elements of a whole complex system or to coupled natural and social systems. In the first case, nature–society relations are internal; in the second case they are external. In either case, however, the crucial question is whether the binding relations represent a weak or a strong coupling. Weakly coupled systems are linear, while strongly coupled systems are non-linear and display complex behaviour.

Returning now to the concept of 'system', it is clear that the meaning of the term SES depends on our understanding of 'system'. As a first approach, we might try using for our construction and analysis of SES conceptual and methodological tools drawn from theories of systems, complexity or graphs. To do this, we need a mathematically oriented definition of the term 'system'. Here we could follow the classical definition of 'system' given by Hall and Fagen (1956), which still finds broad acceptance today among the community of systems thinkers: "A system is a set of objects together with relationships between the objects and between their attributes."

Such a formal definition does not refer to any real object. Instead, systems are defined as sets of related elements, that is, as mathematical objects and classes of abstraction (Becker and Breckling 2011). In research-oriented definitions of 'system', on the other hand, the set-theoretical entities are reinterpreted empirically: Here element is replaced by thing, object, component or part, and relationship by coupling, interaction, binding, connection or linkage. Though helpful for empirical research this remains unsatisfactory from a theoretical point of view, the structure of such empirically interpreted systems continues to be defined mathematically, independent of any semantic interpretation of the elements and relations.

Restrictions on the Formal Definition

Thus the formal definition has serious shortcomings. Mathematically, the notion of 'system' is too general and indistinguishable from 'structured set', 'structure' or 'topological space'. Therefore the definition is incomplete. Two additional restrictions to the general definition are necessary:

1. Definition of the spatial or functional boundaries at different levels
2. Identification of the patterns between the relations, expressed as topological structures (e.g., networks, causal chains, feed-back loops)

Without these restrictions, the notion of a 'system' remains merely a metaphorical way of talking.

General Properties of Systems

Morphologies of special systems have been constructed in General Systems Theory (GST) using pairs of opposing properties (closed/open, static/dynamic, deterministic/stochastic, simple/complex, linear/non-linear, back-coupled/non-back-coupled). Following this guidance, various general properties have been emphasised in SES discourse: open, dynamic, non-linear, back-coupled, complex, adaptive. These general properties have been further condensed in the by now well-known notion of SES as complex and adaptive systems (Berkes et al. 2003).

In this way, the study of complexity has become the central theoretical challenge of SES analysis. In the scholarly network Resilience Alliance, this challenge has been taken up by employing the concept of resilience. In research practice, resilience is conceived more or less as a collection of ideas about how to interpret complex adaptive systems and how to use these interpretations as a source for the generation of hypotheses for concrete case studies. Of course, this is one way of transforming the general boundary object 'SES' into an epistemic object, one strongly structured by the resilience theory.

However, as Janssen et al. (2006) have noted, the work in this direction has so far lacked a clear framework, that is, a clear formal description of structural changes, although the latter is supposed to be one of the key aspects of resilience research. Therefore, they propose a network perspective for the study of resilience in SES. They represent such systems, using the analytical tools of graph theory and network topology as comprising dynamic networks with nodes and links. Nodes may represent not only social actors at different levels, such as individuals, farmers, conservationists, communities, organisations, and so on, but also ecological entities such as arable land, lakes, rivers, forests or paddy fields. The links can symbolise flows of physical units such as water or raw material, and organisms such as seed dispersals or cattle; they can also symbolise the exchange and management of information between social actors.

The concept of SES as networks, as introduced by Janssen et al. (2006), is applied to ecosystems that have been affected by human activities and those have undergone quantitative and qualitative change in both nodes and links, and therefore have also undergone changes in network structure and topology. On the basis of their comparative analysis of several case studies, Janssen et al. (2006) distinguish three archetypical social-ecological networks:

1. Ecosystem networks that are connected by people through the flow of information or materials
2. Ecosystem networks that are disconnected and fragmented by the action of people
3. Artificial ecosystem networks that are created by people, such as irrigation systems or supply systems

Janssen et al.'s (2006) study is one of the rare examples of a formally demanding analysis of SES. However, besides network topology there are several other advanced tools for the formal modelling of complex adaptive systems: non-linear dynamics, adaptive landscapes, multi-agent modelling, neuronal networks, fuzzy logic, genetic algorithm, cellular automata, theory of fractals and chaos. These tools have been forged, tested and applied mostly in the field of complexity studies, but they are slowly trickling into the discourse on human–nature interaction. If we view SES as epistemic objects (having as their components knowledge, problems and methods), the use of these advanced and powerful modelling techniques indicates the possibility of the strong structuring of a formerly weakly structured object.

CONSTRUCTIONS OF SOCIAL-ECOLOGICAL SYSTEMS

The construction of an SES in the strict sense begins with the general properties of such a system mentioned previously, and goes on with a definition of the unit of analysis.

A Working Definition of Social-Ecological Systems

Glaser et al. (2008) and Glaser (this volume) have given a working definition of the concept of SES:

> A social-ecological system consists of a bio-geo-physical unit and its associated social actors and institutions. Social-ecological systems are complex and adaptive and delimited by spatial or functional boundaries surrounding particular ecosystems and their problem context.

We can find similar definitions in the publications issued by the Resilience Alliance (Berkes et al. 2003). It is an open question, however, whether the term 'complex adaptive system' is used there as a formalised analytic-descriptive concept suitable for system analysis using advanced tools, or if it is merely being employed as a narrative and heuristic metaphor for the interpretation of case studies. In any case, boundaries are defined empirically as forming the periphery of a particular unit of analysis, thereby delineating it. Characterizing this unit as a bio-geophysical unit or 'ecosystem' involves a strong operation of structuring since the knowledge, problems and methods implicated in this structuring depend crucially on ecosystem theory. Thus SES are conceptualised mainly with reference to ecosystems affected or managed by human activities.

In my view, the working definition is embedded too deeply in resilience research on managed ecosystems to function as a pragmatic tool for use by a wider range of researchers. The definition of the boundary as boundary surrounding particular ecosystems depends too strongly on the units of ecosystems research. We must, then, in order to make the definition available for a broad spectrum of research activities, first remove this specific determination of the boundary and change the unit of analysis. One should remember here that as long as their boundary is not determined, these units are not systems in the strict sense. Therefore I propose to call such a conceptually undetermined unit of analysis an entity and suggest three types of units, defining three possible levels of analysis:

1. Natural entities in a social context, such as managed local or regional ecosystems—studied in resilience research
2. Social entities in an ecological context, such as climate politics or nature conservation—studied in environmental sociology or ecological economics
3. Hybrid entities, such as humans-in-action, supply systems for water or food in a social and ecological context—studied in human and social ecology

Humans are hybrids in that they are both natural and cultural beings. To introduce human action into ecosystem analysis implies the acceptance of hybrid entities.

To define a unit as a system, the boundary should be defined as surrounding a particular problem context within a unit. The structure and the dynamics of SES depend strongly on their boundary conditions and their problem context. This means that SES analysis is a form of transdisciplinary research (Jahn 2008). Here it is crucial whether the boundary is created and maintained by internal system activities, or is just a demarcation made for the convenience of analysis. In any case, the definition of a spatial or functional boundary introduces empirical conditions into formal systems analysis.

The Reality of Social-Ecological Systems

In Glaser et al.'s (2008) working definition cited earlier, SES are understood to be concrete units in the real world of spatial-temporal phenomena. This is obviously a realistic ontological position. But within the ecological discourse, as well as in other disciplines, realistic ontological assumptions have often been criticised, with this criticism usually employing strong constructivist arguments—SES are, it is argued, only scientific constructions. Various versions of both 'realist' and 'constructivist' positions can be found in ecology and human ecology as well (Becker and Breckling 2011), while within the discourse on SES that centres on the concept of resilience, the controversy is sharpened by adherence to a hardcore empiricism, leaving realistic positions de facto dominant, with epistemological discussions of alternatives being the exception.

Now any strict definition of the term 'system' will hold that, from an epistemic point of view, systems are mathematical objects—that is, classes of abstraction. At the same time, we have to recognise that there exist social-ecological phenomena in the real world. Therefore I advocate a model-oriented constructivist realism that combines a realistic ontology with a constructivist epistemology. From this point of view SES appear as models of knowledge about real-world phenomena.

As with any realistic ontology, this position is in danger of falling prey to the "fallacy of misplaced concreteness" as analysed by Alfred North Whitehead (1929: 20): "The 'fallacy of misplaced concreteness' consists in neglecting the degree of abstraction involved when an actual entity is considered merely so far as it exemplifies certain categories of thought." The term 'complex SES' is a concept, a category of thought, an abstraction from reality, and not an actual entity. The danger of the fallacy of misplaced concreteness arises whenever this category of thought is confused with a real object like an oasis in the Sahara desert—when, for example, the empirical properties of the 'oasis' are attributed without conceptual restrictions to the formal attributes of 'complex SES'.

Constructivist realism is only meaningful if we distinguish between the 'real world' of concrete things and processes in space and time and an 'ideal world' of abstract objects. Such abstract objects may be logical-mathematical in nature (equations, mathematical spaces, topologies, etc.), capable of being modelled, at least in principle, by a computer program. They may also be verbal, graphic, metaphorical or conceptual descriptions, which aim at representing knowledge about complex networks of interaction. We can define systems, then, as abstract objects in an ideal world.

Constructivist realism concentrates its attention on relationships between the ideal and the real spheres. To do so it posits a model relation between systems as abstract objects and concrete real-world phenomena. Therefore system—an abstract object—can serve as a model of real-world social-ecological phenomena[5]. However, now the question about the relationship between model and real-world phenomena arises.

The Construction Process of Social-Ecological Systems

SES are constructed within an "epistemic circle" (see Tretter and Halliday, this volume) that connects several processes:

- The process starts with a selection of the empirical unit of analysis. Looking at different areas of research, we have distinguished three possible types of units: natural, social and hybrid entities. These types correspond roughly to the natural, social and human ecological fields of research.
- In the next phase, an abstraction is necessary from empirical observations of real-world entities and their contingent properties, and also from the context of the unit of analysis.
- For the construction of SES as idealised objects, a constitutive distinction between nature and society (or nature and culture) at different levels (element, system, subsystem, super-system) is necessary.

- Finally, in a move back towards concretization, an interpretation of the abstract system in empirical terms leads to the construction of a model for the unit in consideration. Elements and relations referring to real-world phenomena have to be identified, spatial or functional boundaries at different levels must be defined, and variables that indicate system properties have to be found.

Mental Models as Generator of Epistemic Objects

The transformation of a vaguely defined boundary object into a strongly structured epistemic object is guided, explicitly or implicitly, by pre-analytic ideas, general worldviews and ontological convictions. These function as mental models of the part of the world, together with its major problems we are interested in, and generate images of relevant issues or processes. Marion Glaser (2006) has introduced the term 'mind map' for these pre-analytic visualizations. Examining examples of mind maps of human–nature relations with reference to the social dimension in ecosystem management, she distinguishes four major kinds of mind maps (with several sub-types): the ecocentric, the anthropocentric, the interdisciplinary and the complex. Here, however, I wish for a moment to look at another question: How does the boundary object 'SES' get transformed into an epistemic object? And in the process of transforming a boundary object into an epistemic object, what role do mental models play?

Mind maps of nature–society interactions are created by a constitutive distinction between 'nature' and 'society' (or 'nature' and 'culture'). Normally, the topology of such maps is symbolised graphically by simple set diagrams or schemes of ontological levels (physical, chemical, biological, human, social, spiritual, etc.). And normally the question is left open up to which ontological level the distinction is being carried out. 'Nature' or 'society' may symbolise either a unified set of 'natural' entities (physical, chemical, biological, geological) or 'societal' entities (values, institutions, knowledge, social networks), respectively—or all these entities bound together in a hybrid system.

ANALYSIS OF A SIMPLIFIED ARTIFICIAL WORLD

We have introduced above three types of units, defining three possible levels of analysis: natural, social and hybrid entities. In case studies about delimited real-world phenomena, these levels of analysis represent not only three different aspects of an SES, but also three different epistemological objects with different structures. In other words, selecting a level of analysis is a first step of transforming a boundary object into an epistemic object. In order to prove this argument, I have invented a simplified relation-oriented artificial world[6] as an example of demonstration.

Normally our perception of the world emphasises permanent things with distinctive qualities. In contrast, in a relation-oriented world view (Bateson 1972) things are never isolated; they are coagulated patterns, hardened into more or less durable 'objects' at the crossing-points of flows of matter, energies and information. In ecosystems or in human societies with living and perceiving beings, many different kinds of activities generate dynamic relationships between distinct things, effecting weak or strong bindings, with new relations continually arising and old ones constantly disappearing—and along with the latter many things as well. Transformations of patterns and structures are going on continuously.

A Simple World as Social-Ecological Network

In our normal perception of the world, we name distinguished objects, thus giving them a meaning within the universe of language, as well as bestowing on them an ontological status. However, "the map is not the territory, and the name is not the named thing", as Gregory Bateson (1979) noted. In Figure 3.2, several identified objects are named with numbers, with any number capable of representing a different thing. If we think of a social-

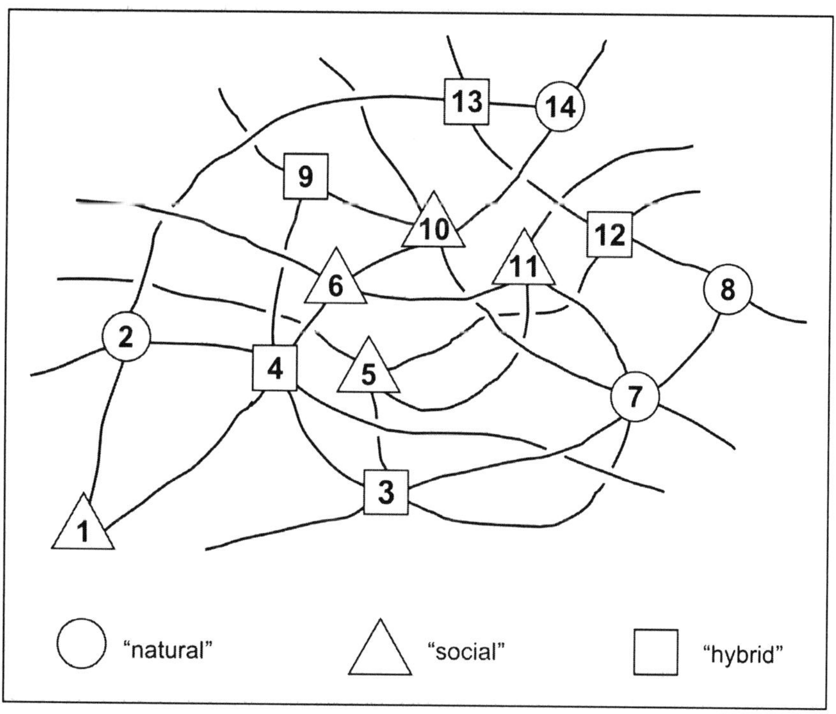

Figure 3.2 A simple world as social-ecological network.

ecological unit such as a managed forest, these objects may represent plants and trees, foresters and woodcutters, deer and wild boars, geophysical objects, or human-made artefacts like wells—whatever seems to be relevant for an analysis of the forest in consideration.

With the idea of an SES as boundary object, the different objects under consideration have to be classified either as 'natural' or 'social'. This has been done arbitrarily in Figure 3.2. Some objects, in particular humans-in-action, are both natural and social; they are categorised as 'hybrid'. Only the qualities of objects so marked are significant for further consideration. These objects make up the elements of the SES. However, for the construction of a system we also need relations between them. In research practice, these relations may be the flow of matter, energies and information in an ecosystem, water or raw material within a landscape, the traces of living beings, or the flow of information between social actors. Step by step, the decisions about the relevance of particular elements and relations for a problem under consideration transform the general boundary object SES into a social-ecological network, that is, an epistemic object.

The network that emerges from this analysis is a topological object—a mathematical structure in an ideal world of abstractions which may be analysed using the tools of network or graph theory[7] (Janssen et al. 2006). It is also possible to consider the network as the starting point for a multi-agent analysis. In the latter case, the relations sketched in Figure 3.2 would be the result of rule-based actions at a selected moment. Like a snapshot, the graphic representation captures ties or relations, both material and symbolic, including relations of perception.

Creating Mind Maps by Set-Theoretical Operations

Applying the set-theoretical operation 'intersection' with the elements of an SES, and neglecting the relations between them, the whole set is decomposed into subsets, with each class of elements forming a particular subset. It turns out that the subset of all the hybrid elements is located in the intersection of the 'natural' and the 'social' subsets. Obviously, humans-in-action are represented in this subset. This is the abstract form of a mind map of human–nature interaction (Glaser 2006) that is very popular in human and social ecology.

The mind map sketched in Figure 3.3 illustrates that, even at a very high level of abstraction, there is a hybrid zone where nature and society intersect. This is the starting point, for example, for the mind maps used and elaborated on by Vienna social ecology, where human populations are located in the hybrid domain (Fischer-Kowalski and Weisz 1999). As a next step, relations between the components are incorporated into the model, based on the interpretation of real interactions observed between the cultural, natural and hybrid domains. But here too, the fallacy of misplaced concreteness seems hard to avoid.

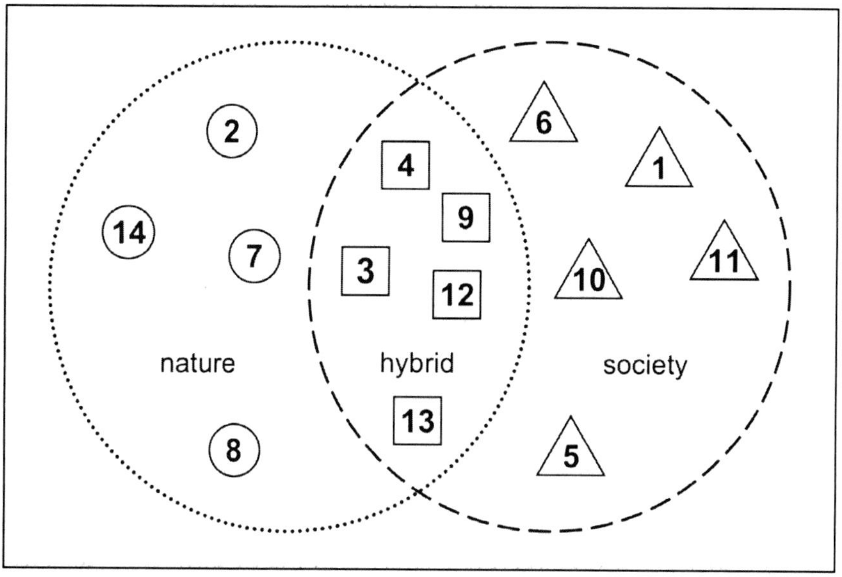

Figure 3.3 Intersection of the 'natural' and the 'social' subsets.

The concept of social-ecological supply systems, as used in Frankfurt social ecology, also employs a mind map of intersecting sets. Within the supply systems, users and resources are distinguished, and several factors (knowledge, practice, institutions, technology) affecting users and resources are introduced (Hummel 2008, Hummel, this volume). Here, the mind map remains unequivocally a conceptual model and therefore only a first step in an empirical analysis of concrete supply systems.

Three Approaches to an Analysis of Social-Ecological Systems

The mind map of Figure 3.3 suggests three possible units of analysis: the natural, the hybrid and the social domain—as already distinguished in the previous section. However, which relations should be taken into consideration, and how should we form the boundary of the selected unit? If we select, for instance, the natural domain as unit of analysis and compare it with the complete network of Figure 3.2, we see that the elements of the natural domain have relations both to the hybrid and to the social domain[8]. Whether or not the selected network shows characteristics of a complex system depends, on the one hand, on the patterns of relationships identified and selected and, on the other, on the borders surrounding the selected elements.

How the boundary is defined turns out to be crucial. We may include the hybrid elements into the natural domain—or not; we may include all social elements that bridge between natural elements—or not. Depending on the form of the boundary, the internal elements and relations are defined, and together with this definition also the structure of the network in consideration, for instance, positive or negative feedback loops relevant for the dynamic of the system and its resilience. With those decisions we arrive at different networks with different structures. We can call all these structures, somewhat fuzzily, 'natural' social-ecological networks. SES system theorists prefer this type of reconstruction, as does the resilience community. The aforementioned three archetypical social-ecological networks (Janssen et al. 2006) are examples of 'natural' social-ecological networks.

We can reconstruct a 'social' and a 'hybrid' social-ecological network in a similar manner. The result is indisputable: The three approaches obviously represent not only three different aspects of the social-ecological network in consideration, but also three different epistemological objects[9].

Which of the three SES networks we use depends on the issue in question and the context of problems. If we leave the simplified world of abstraction and look at real world social-ecological phenomena, three cases present themselves:

1. For the management of ecosystems the 'natural' SES network should be the object of study.
2. If the focus of interest is on the analysis and the management of supply systems or societal metabolism, the 'hybrid' SES network becomes the preferred object.
3. For many studies of sustainable development the 'social' SES network is the appropriate object.

Notice that every network in our simplified world has natural, social and hybrid elements and relations inside its boundary affecting its dynamics. Strictly speaking, each of them is a 'hybrid' SES network with anthropogenic problems within its domain. The three cases distinguished here are conceptually related to (1) ecosystem theory, (2) human and social ecology and (3) environmental sociology or economics, respectively. From the point of view of a theory of societal relations to nature, the hybrid network (2) is the preferred unit of analysis; the two other appear as reductions.

NOTES

1. The concept of societal relations to nature is explained briefly in section 3 of the present chapter.
2. Other distinctions such as those between human and nature or between nature and culture can be derived from this basic distinction between nature and society.
3. Ecosystem services are the benefits people obtain from ecosystems. These include provisioning services such as food and water; regulating services such as flood and disease control; cultural services such as spiritual, recreational,

and cultural benefits; and supporting services such as nutrient cycles that maintain the conditions for life on Earth (Daily 2000; MEA 2005).

4. See, for example, the articles by Beate Ratter as well as Felix Tretter and Andrew Halliday in this volume.
5. 'Social-ecological phenomena' are concrete associations of entities in a real world. We call them *social-ecological compounds* (or *complexes*)—and distinguish them conceptually from *SES* as models.
6. The graphic representation has no reference to any real things. The different lines and their intersections were drawn arbitrarily. An intersection-free graph of a two-dimensional web would require three dimensions. To avoid a three-dimensional graph, the lines representing the relations are interrupted if they do not intersect.
7. A powerful tool for such an analysis is an adjacent matrix: It is possible to condense the entire information contained in the network graph into such a matrix (Tittmann 2003; Berge 2001).
8. Using the adjacent matrix mentioned before as a tool we can analyse the structure of the natural domain in a simple way.
9. It is possible to integrate the different results into one structure by using the tools of network topology, particularly the adjacent matrix mentioned already. The three networks are then seen to be overlapping sub-matrices.

REFERENCES

Anderies, JM, Walker, BH, Kinzig, AP (2006) Fifteen weddings and a funeral: Case studies and resilience-based management. Ecology and Society 11: Art. 21 (online)

Bateson, G (1979) *Mind and Nature: A Necessary Unit.* E. P. Dutton, New York

Bateson, G (1972) *Steps to an Ecology of Mind.* University of Chicago Press, Chicago, London

Beck, U (1992) *Risk Society: Towards a New Modernity.* Sage, London

Beck, U (1986) *Risikogesellschaft. Auf dem Weg in eine andere Moderne.* Suhrkamp, Frankfurt am Main

Becker, E (2006) Problem transformations in transdisciplinary research. In: Hirsch-Hadorn, G (ed) *Unity of Knowledge.* In: Encyclopedia of Life Support Systems (EOLSS), Oxford, UK (online: http//:eolss.net)

Becker, E, Jahn, T (2006): *Soziale Ökologie. Grundzüge einer Wissenschaft von den gesellschaftlichen Naturverhältnissen.* Campus, Frankfurt am Main, New York

Becker, E, Jahn, T (2003) Societal relations to nature: Outline of a critical theory in the ecological crisis. (online: www.isoe.de)

Becker, E, Breckling, B (2011) Border zones between ecology and systems theory. In: Schwarz, AE, Jax, K (eds) *Revisiting Ecology: Reflecting Concept, Advancing Science.* Springer, Berlin

Berge, C (2001) *The Theory of Graphs.* Dover, New York

Berkes, F, Colding, J, Folke, C (2003) *Navigating Social Ecological Systems: Building Resilience for Complexity and Change.* Cambridge University Press, Cambridge

Brand, FS, Jax, K (2007) Focusing the meaning(s) of resilience: Resilience as a descriptive concept and a boundary object. Ecology and Society 12: Art. 23 (online)

Crutzen, PJ. (2002) Geology of mankind. Nature 415: 23

Daily, GC (2000) Management objectives for the protection of ecosystem services. Environmental Science & Policy 3: 333–339

Dürr, HP, Oesterreicher, M (2001) *Wir erleben mehr als wir begreifen. Quantenphysik und Lebensfragen.* Herder, Freiburg Basel Wien
Fischer-Kowalski, M, Weisz, H (1999) Society as hybrid between material and symbolic realms: Towards a theoretical framework of society-nature interaction. Advances in Human Ecology 8: 215–251
Folke, C (2006) Resilience: The emergence of a perspective for social-ecological systems analysis. Global Environmental Change 16: 253–267
Galison, P (1991) The trading zone: Coordinating action and belief in modern physics. Conference at the Institute Henri Poincaré. Paris, 13. December 1991
Glaser, M (2006) The social dimension in ecosystem management: Strength and weaknesses of human–nature mind maps. Human Ecology Review 13: 122–142
Glaser, M, Krause, G, Ratter, B, Welp, M (2008) Human–nature interaction in the Anthropocene: Potential of social-ecological systems. GAIA 1/08: 77–80.
Habermas, J (2005) *Zwischen Naturalismus und Religion.* Suhrkamp, Frankfurt am Main
Hall, AD, Fagen, RE (1956) Systems engineering. Vol. I. New York (Bell Telephone Laboratories). In: Buckley, W (ed) (1968) *Modern Systems Research for the Behavioral Scientist.* Aldine, Chicago, pp 81–92
Hummel, D (2008) *Population Dynamics and Supply Systems: A Transdisciplinary Approach.* Campus, Frankfurt, New York
Intergovernmental Panel on Climate Change (IPCC) (2007) IPCC Fourth Assessment Report. Cambridge University Press, Cambridge
Janssen, MA, Bodin, Ö, Anderies, JM, Elmquist, T, Ernstson, H, McAllister, RR, Olsson, P, Ryan, P (2006) Toward a network perspective of the study of resilience in social-ecological systems. Ecology and Society 11: Art. 15 (online)
Jahn, T (2008) Transdisziplinarität in der Forschungspraxis. In: Bergmann, M, Schramm, E (eds) *Transdisziplinäre Forschung.* Campus, Frankfurt am Main, New York, pp 21–37
Kates, RW, Clark, WC, Corell, R, Hall, JM, Jaeger, CC, Lowe, I, McCarthy, JJ, Schellnhuber, HJ, Bolin, B, Dickson, NM, Faucheux, S, Gallopin, GC, Grübler, A, Huntley, B, Jäger, J, Jodha, NS, Kasperson, RE, Mabogunje, A, Matson, P, Mooney, H, Moore III, B, O'Riordan, T, Svedin, U (2001) Sustainability science. Science 292: 641–642
Latour, B (2005) *Reassembling the Social: An Introduction to Actor-Network-Theory.* Oxford University Press, Oxford
Liu, J, Dietz, T, Carpenter, SR, Alberti, M, Folke, C, Moran, E, Pell, AN, Deadman, P, Kratz, T, Lubchenco, J, Ostrom, E, Ouyang, Z, Provencher, W, Redman, CL, Schneiderand, SH, Taylor, WW (2007) Complexity of coupled human and natural systems. Science 317: 1513–1516
Luhmann, N (1995) *Social Systems.* Stanford University Press, Stanford, CA
Marx, K (1857) *Grundrisse der Kritik der Politischen Ökonomie (Rohentwurf).* Dietz Verlag, Berlin
Millennium Ecosystem Assessment (MEA) (2005) *Ecosystems and Human Well-Being: Synthesis.* Island Press, Washington, DC
Rheinberger, HJ (1997) *Toward a History of Epistemic Things: Synthesizing Proteins in the Test Tube.* Stanford University Press, Stanford, CA
Rheinberger, HJ (2006) *Epistemologie des Konkreten. Studien zur Geschichte der modernen Biologie.* Suhrkamp, Frankfurt am Main
Schellnhuber, HJ, Crutzen, PJ, Clark, WC, Clausen, M, Held, H (2004) *Earth System Analysis for Sustainability.* MIT Press, Cambridge, MA
Star, SL, Griesemer, JR (1989) Institutional ecology, 'translations' and boundary objects: Amateurs and professionals in Berkeley's Museum of Vertebrate Zoology, 1907–39. Social Studies in Science 19: 387–420

Tittmann, P (2003) *Graphentheorie. Eine anwendungsorientierte Einführung.* Fachbuchverlag Leipzig, Leipzig

Vitousek, PM, Mooney, HA, Lubchenco, J, Melillo, JM (1997) Human domination of Earth's ecosystems. Science 277: 494–499

Whitehead, AN (1978) *Process and Reality: An Essay in Cosmology.* The Free Press, New York

Zalasiewicz, J, Williams, M, Smith, A, Barry, TL, Coe, AL, Bown, PR, Brenchley, P, Cantrill, D, Gale, A, Gibbard, P, Gregory, FJ, Hounslow, MW, Kerr, AC, Pearson, P, Knox, R, Powell, J, Waters, C, Marshall, J, Oates, M, Rawson, P, Stone, P (2008) Are we now living in the Anthropocene? In: GSA Today. The Geological Society of America 18: 1–8 (online)

4 Modelling Social-Ecological Systems

Bridging the Gap Between Natural and Social Sciences

Felix Tretter and Andrew Halliday

Human ecology without systems science is blind and systems science without human ecology is empty. (Tretter 1998)

INTRODUCTION: WHY MODEL SOCIAL-ECOLOGICAL SYSTEMS?

The management of current environmental problems requires an integrative understanding of social-ecological systems. Various systems approaches can help us make sense of, and respond to, the multiple interdependencies between social and ecological dimensions of our world (Tretter 1979, 1982; Bühl 1990; Bossel 2004; Forrester 1971; Liu et al. 2007; Odum 2007). Models of social-ecological systems can help human ecology achieve its aim of explaining the co-evolution of society and nature—and provide guidelines for systemic societal intervention to mitigate environmental problems such as climate change.

To this end we can draw on a number of different strands in systems thinking, from early work by the founders of cybernetics (Wiener 1948) and General Systems Theory (GST; Bertalanffy 1968) into the dynamics of biological systems to recent advances in agent-based modelling of social and biological systems (Deaton and Winebrake 2000; Hill and Coquillard 2007; Janssen and Carpenter 1999; Janssen et al. 2006; Janssen and Ostrom 2006; Sloot and Hoekstra 2007; Watson and Lovelock 1983; Wilensky 2001; see also the chapters by Ratter, Glaser and Wildenberg and Singh, this volume). We can apply concepts from cybernetics by building a descriptive functional model, which proceeds from the initial attribution of certain dynamic properties to nature and society to explore the mutual dependencies that exist between them, with the aim of identifying significant factors that can be socially controlled and could have a damping effect on environmental problems. 'Running' the model will enable us to evaluate different societal strategies which could be adopted to manage human–nature dynamics and ensure the sustainability of social-ecological systems and thereby also of the livelihoods of future generations.

It goes without saying that there are considerable analytical and practical challenges involved if systems models are to be constructed. These derive fundamentally from the fact that nature and society themselves can be seen as complex self-regulating networks (Figure 4.1). Thus, in order to specify appropriate control actions for the management of social-ecological systems, detailed information is required on

- The state of the natural environment—and the validity of the data we use to measure it;
- The speed and scale of nature's response to human disturbance
- The robustness of natural systems—and its limits—in the face of human induced environmental change
- The feedback mechanisms which enable social systems to perceive, conceptualise and respond to ecological problems
- The nature of this response, through changes in behaviour of critical subsystems (e.g., science, industry, politics, law) and subpopulations (e.g., consumers, farmers, tourists, industrialists)

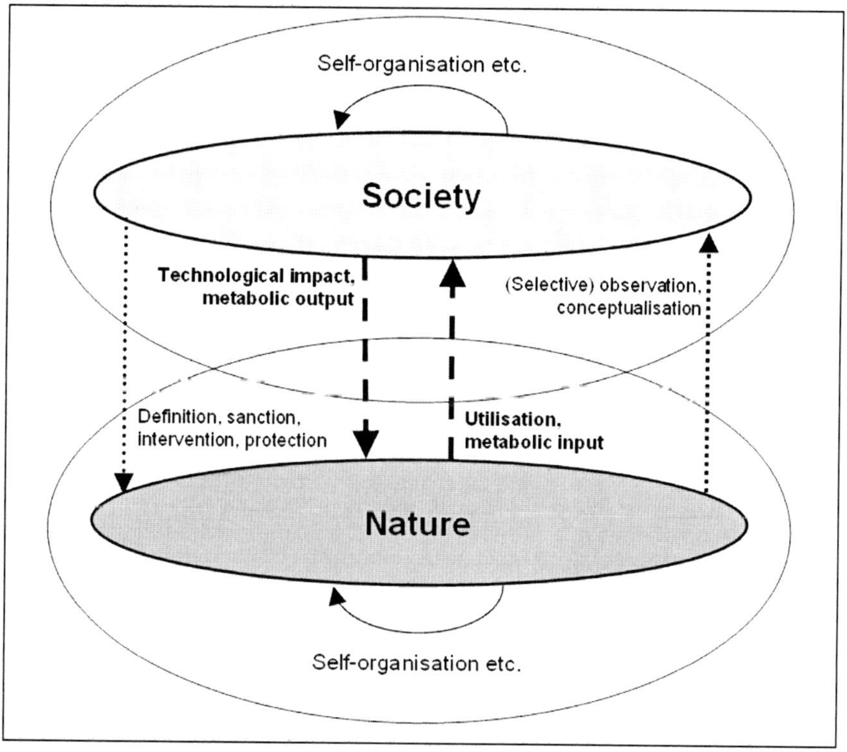

Figure 4.1 The functional structure of the social-ecological control problem. The natural system is utilised by the social system, the actions of the social system influence—positively or negatively—the natural system. Both the natural system and the social system have intrinsic dynamic properties (self-organisation, slow and fast processes, non-linearities) that do not allow the application of simple input-output or stimulus-response models.

In setting out to represent these and other variables and their interdependencies in the form of a conceptual model, a number of methodological difficulties arise:

- The extreme complexity of the processes involved makes it impossible to come close to a complete model of the system. So the question arises, how many components should be considered in modelling complex social-ecological systems? Which should be included and which should be left out?
- The data we need is often unavailable; moreover available data sets are often incompatible for the purposes of integrative analysis.
- More generally, the methodological gap between natural sciences and social sciences makes the construction of a single integrative model very difficult.

We still lack generally applicable theoretical laws to explain the behaviour of real complex dynamic systems.

These challenges and ways to meet them form the subject matter of this chapter: The next section discusses specific philosophical, theoretical and methodological issues that need to be taken into account in modelling social-ecological systems. Then we move on to a consideration of some key concepts in human ecology. This is followed by a brief review of the basic procedures involved in quantitative modelling. The final section provides an overview of currently available methods and techniques for modelling social-ecological systems from a systems science perspective.

PHILOSOPHICAL, THEORETICAL AND METHODOLOGICAL CONSIDERATIONS

In setting out to construct a model of a social-ecological system it is necessary to take into consideration a number of theoretical and methodological issues related to wider questions in the philosophy of science that are usually not considered explicitly. These are outlined briefly in the following paragraphs.

The Functions and Limitations of Modelling

The American statistician George Box (1979) proposed that "all models are wrong; some are useful".

The truth of the first half of this statement can be appreciated when we consider that modelling proceeds, first of all, through a process of mapping, by which features x_1–x_n in the situation to be modelled are assigned characteristics y_1–y_n in the system model (Holland 1998). Thus the model is both a mapping representation and construct produced by the observer (Stachowiak 1973); it is "constructivist realism" (Becker, this volume).

Understanding this procedure permits an appreciation of both the limitations and usefulness of modelling. In the words of Alfred Korzybski (1994), "[A] map is not the territory it represents, but, if correct, it has a similar structure to the territory, which accounts for its usefulness . . . If the structure is not similar, then the traveller is led astray."

From the standpoint of complex systems, the epistemic function of modelling is thus as a heuristic strategy for understanding processes that occur in these systems (Ossimitz 2000; Richmond 2001; Érdi 2008). Despite their intrinsic uncertainties, models are the only way we have to forecast ranges of possible behaviour of complex systems. However, Korzybski's metaphor of the traveller is a useful one because it reminds us that models cannot *explain* the behaviour of complex systems in the real world. As mental constructs they help us *explore* their behaviour, by demonstrating how the model behaves under certain conditions assigned by the modeller, which are believed to have a similar structure to past, present or future conditions in the real world.

To the extent that models incorporate theories that have been developed to explain the behaviour of complex systems, they can be used to test these theories, by comparing the behaviour of the model with what happens in the real world. From the standpoint of human ecology, however, three different types of theories have to be integrated into models of social-ecological systems: theories about the behaviour and self-organisation of (1) natural systems and (2) social systems; and (3) universal theories of systems dynamics. This indicates the scale of the challenges confronting would-be modellers of social-ecological systems.

Objective vs. Non-Objective Knowledge

The study of social-ecological systems straddles the border between the natural and social sciences. While the possibility of 'objective knowledge' in natural science is generally accepted; in the case of the social sciences, it is often maintained that that the impossibility of the scientific observer standing outside the object of study limits the objectivity of acquired knowledge or even makes objective knowledge impossible altogether. Natural science aims to explain phenomena, while social science aims to interpret and understand them in the hermeneutic tradition (Giddens 1976; Checkland [1981] 1984; von Wright 2004).

This brief chapter cannot of course go fully into the wide-ranging and complex debates among philosophers of science around this question. Nor would it be fruitful to do so. Key issues for the study of social-ecological systems relate to (1) the reality of ecological problems, (2) the nature of our knowledge about them and (3) the methodological approach to adopt in studying these problems from the perspective of human ecology.

From a radical constructivist perspective it is possible to deny the reality of ecological problems. In this view, ecological problems only exist for society if they are communicated by observers and defined by them as problems for society. There is no 'real' problem; the natural environment itself is also a social construct. The environmental sociologists Catton and Dunlap (Dunlap 1979; Dunlap and Catton 1994) drew attention to the paradigmatic assumptions underlying these philosophical speculations and their influence on social science. These assumptions are incompatible with the standpoint of human ecology, which recognises the reality of the natural environment, and acknowledges that human beings, for all their distinctive features, are dependent on their natural environment and subject to environmental constraints just like any other species. These rather abstract epistemological debates acquire practical significance when choices have to be made about the methodological approach to adopt in modelling social-ecological systems.

Natural scientists consider that natural systems consist of bio-geophysical components and system models focus on flows of matter and energy (Odum 2007). Models of social systems focus on patterns of human action or are conceived as systems of communication, that is, flows of meaning (Luhmann 1984, 1986; Luhmann et al. 1996). While in principle it is easy to see that a model of a social-ecological system should cover all these aspects, in practice this is complicated by the difficulty—or impossibility—of finding a common language and conceptual framework to describe these phenomena: matter and meaning are mutually irreducible to each other (Halliday 2005).

A number of attempts have been made to integrate sociological, ecological and systemic thinking (notably by Bühl 1990). Nevertheless, the preceding considerations suggest that it may be impossible to develop a complete model of a social-ecological system, even without considering practical problems relating to volumes of data and system complexity. At the outset a decision is required on what kinds of knowledge to include in the system model, and what kinds to leave out. The decision taken will depend on the research interests of the modellers, the purpose of the model and the kind of data the model is intended to generate.

The Use of Qualitative and Quantitative Data

In the mapping procedure described already, the features of reality we wish to incorporate into our model may be represented by both quantitative and qualitative data. Quantitative modelling becomes possible if, in the mapping process, we are able to transform all data into quantitative values that are amenable, for example, to computer-based simulation. Knowledge in the social sciences is mainly formulated in the form of verbal, that is, qualitative models (Luhmann 1984), and this is also generally the case for human ecology. Qualitative conceptual frameworks can be transformed into quantitative models by developing procedures to assign numbers to qualitative values. We take part in this process every time we answer a questionnaire that, for example, asks us to give our opinion about a statement on a scale of 1 ('strongly disagree') to 5 ('strongly agree'). Few things are inherently unquantifiable; as Sterman (2002) notes, "[H]uman creativity is great: once we recognise the importance of a concept, we can almost always find ways to measure it."

The natural sciences, on the other hand, have access to a large pool of quantitative data that provides the raw material for mathematical models. But this data also requires processing—selecting and simplifying—to make it suitable for incorporation into quantitative models. This procedure involves qualitative analysis, regarding for example the relative importance or representativity of different kinds of data. The selection of data for inclusion in the model will draw on qualitative theoretical models. Ethical considerations and the value positions held by the modellers will also come into play, as well as operational considerations related to the availability of the data.

For example, models of global systems that work with economic values commonly include a 'discount rate' which implies a downward adjustment in the value that we place today on the production of goods and services in the future. Deciding on the discount rate involves both theoretical speculation about likely long-term economic trends and ethical questions regarding the importance we attach to the well-being of future generations, compared to our own (Boumans et al. 2002). Thus the modelling process involves us in an 'epistemic cycle', from empirical observations to theory and back and from quality to quantity and back (See Figure 4.2; Tretter 2005, comp. Bunge 1998).

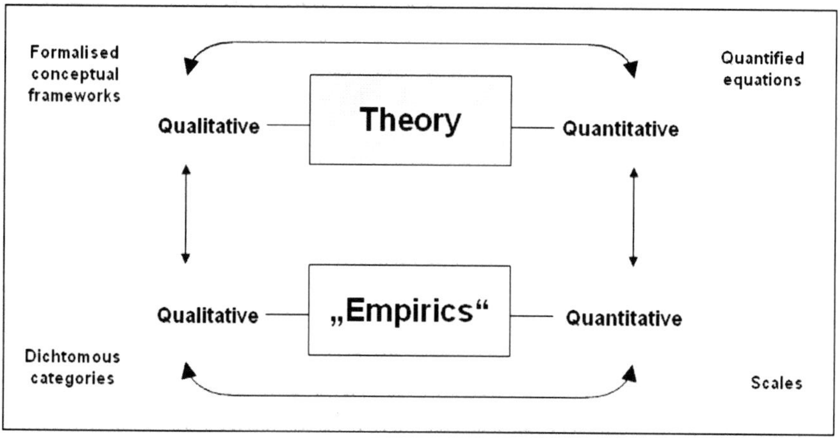

Figure 4.2 The 'epistemic cycle'—from empirical observations to theory and back, and from quality to quantity and back. Human ecological knowledge is mainly formulated in verbal models, whereas systems science generates formal models that use quantitative data from empirical research.

BOUNDARY DRAWING: WHAT TO INCLUDE IN THE MODEL

As noted previously, when building a model of a social-ecological system, one must start out by defining the elements of the model and the relations between them. This is a boundary drawing process: At the same time we are deciding which elements and relations to leave outside the boundary and not to include in our model. All the theoretical considerations considered in the preceding sections can be formulated in the form of boundary drawing questions. In fact it can be argued that boundary judgements are the central issue in modelling complex systems, not only first-order boundary judgements about what is inside and outside the system, but also second-order boundary judgements about what it is that gives rise to boundary judgement in the first place (e.g., who decides and what criteria are used; Midgley 2000, see also Halliday and Glaser 2011).

Among the most important first-order boundary judgements for social-ecological systems are those which determine the spatiotemporal boundaries of the system of interest. Social-ecological systems exhibit wide-ranging connectivity across and between spatial scales, with global forces, such as the demand for natural resources, playing a key role in the dynamics of local systems. Nevertheless a decision has to be made as to which elements are to be considered as part of the internal workings of the system and which as external forces acting on the system. That is, we have to determine the principal spatial and temporal scales of interest for our study and draw a geographical boundary around the social-ecological system to be analysed. At this stage the problem may well arise that it is difficult to find an appropriate fit between the geographical boundaries used for the analysis of social and ecological systems (Folke et al. 1998; Cash et al. 2006). The choice will be determined by the research interest, as well as by practical and operational considerations, such as the availability of data.

For any given geographical area there will be a vast range of social and ecological issues that could be considered. Boundary drawing is also a process of selecting the issues that we wish to address in the design of the model; this in turn guides the selection of components, variables and sources of data. As discussed below, three different perspectives or types of model are possible, corresponding to different research interests: biocentric, sociocentric and hybrid (see also Becker, this volume). This implies corresponding boundary judgements, for example, a decision to place broader social issues outside the system boundary in biocentric models. In hybrid models, practical considerations may well entail focusing on elements of importance for human–nature relations, leaving many 'internal' social and ecological dynamics outside the system boundary.

COMBINING DATA FROM NATURAL SCIENCES AND SOCIAL SCIENCES

A further challenge which arises in the formal modelling of social-ecological systems is to integrate 'hard' data of physical sciences with 'soft' data of social sciences into single 'working model' of the system. For example, concepts such as 'energy' and 'power' can be measured quite precisely in physical systems. Using Joules as universally accepted measurement units, detailed dynamical relations can be built into the model to represent physical energy flows in the system and changing patterns of human resource use (Odum 1994, 2007; Montague 2007). Corresponding concepts in the social sciences, such as 'political power' or 'empowerment', for example, are much harder to operationalise in this way, even when procedures exist for expressing them in quantitative terms. Thus it appears that social drivers of social-ecological systems can be incorporated into quantitative models only at much coarser levels of definition, compared to physical drivers, for example, by considering society as a homogeneous entity and ignoring social differences (e.g., see Boumans et al. 2002; Wildenberg and Singh, this volume). Alternatively, formal modelling of physical aspects of the system can be used to inform a qualitative analysis of the interrelation between physical and social variables in the system (e.g., see Campbell 2004).

In more general terms, the dynamical relations we build into our model will derive from our interpretation of comparative data and time series data, which we use to postulate relations between variables. Such interpretation is rarely straightforward and statistical tools such as factor analysis, cluster analysis, path analysis, time series analysis, and logistic regression analysis are required to validate trends observed in samples of data points (Box et al. 2008). Such statistical procedures are essential to inform discussions of, for example, data that appears to indicate that the global climate will become warmer and that this change is caused by human activities. The challenge is then to identify equations (e.g., differential equations) that describe the relations between the variables under consideration and give rise to the 'line of best fit' that has been validated by statistical procedures. Such equations can be considered as representing mechanisms at work in the system that generate observed behaviour, and enable us to make predictions about the future behaviour of the system. However, the identification of a past or current trend does not provide information about whether or not the trend will continue. The assumptions built into the model about probable future trends will also depend on more general theoretical consideration, for example, whether ecosystem dynamics is considered to follow a pattern of growth leading to a climactic steady state configuration, or to follow a pulsating, cyclical pattern of change (Campbell 2004; Holling and Gunderson 2002).

Classifying System Behaviour: How to Know What to Look For

In interpreting the results of running computer-based models of complex social-ecological systems, attention will often focus on the emergent properties resulting from different patterns of relations, such as sustainability, resilience, vulnerability, adaptivity and transformability. However, even though such terms are well established in ecology, as qualitative descriptions of dynamic properties of self-organising systems, there is still no consensus on their formal (i.e., mathematical) meaning in systems science. To be useful for systems analysis, these qualitative descriptions of system functionality and behaviour need to be transformed into formal functional terms. Assuming that the model is based around stimulus-response relations (black box model), one way to do this is to look for evidence of these patterns in the time courses (trajectories) of state variables (Holling and Gunderson 2002; Walker et al. 2004; Tretter 2005).

For example, the term 'resilience' has been defined in the context of social-ecological system research as "the capacity of the system to absorb disturbance and reorganise while undergoing change so as to retain essentially the same function, structure, identify and feedbacks" (Walker et al. 2004). This definition of 'ecological resilience' expands on the more restricted traditional definition of 'engineering resilience' as the tendency of the system to return to its initial state after a perturbation. The wider definition encompasses resilience in this narrow sense but is also conceptually related to other system theoretical terms such as resistance, stability, robustness, persistence and inertia. Figure 4.3 shows a hypothetical example of the response of a state variable to a disturbance at time t=0. The latency of

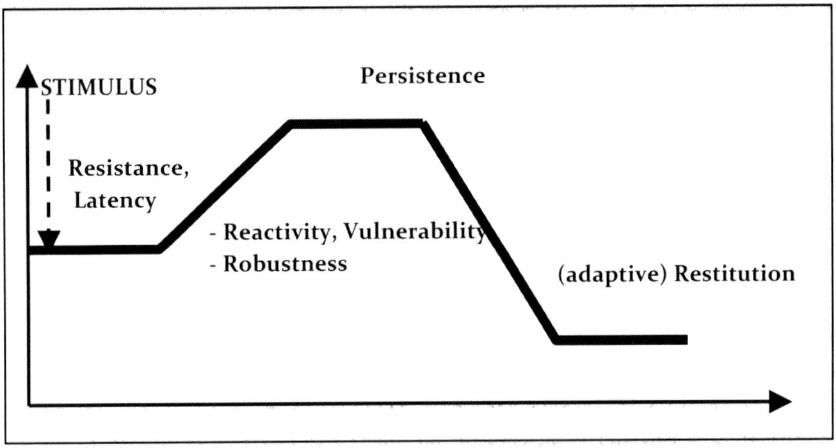

Figure 4.3 Time course of a system's reactions to a stimulus in terms of input-output analysis and qualitative characteristics of the curve.

the response is an indicator of resistance. The subsequent increased value of the state variable in response to the disturbance remains within specified vulnerability limits; thus is it possible for this altered state to persist over time without further perturbation. This pattern shows robustness, that is, the ability of the system to continue to function under changed conditions, and to persist over time. This persistence could be the result of a number of adaptive mechanisms by the system, for example, the shutting down of non-essential activities during a period of stress, or self-organised storage processes that absorb inputs in ways that reduce their impact on system functionality. This is followed by a period of restitution, showing the resilience of the system in the narrower sense of engineering resilience, that is, the return of the system to its former state. 'Adaptive restitution' would indicate that the system has responded to the disturbance by shifting to a lower value state to prepare for possible future perturbations.

The foregoing discussion relates to resilience in the response of a single state variable to a disturbance. But clearly, more information is needed to assess the resilience of the entire system. A more complete picture could be obtained by ascertaining the resilience of a set of key state variables judged to be essential for the continued functionality and identity of the system. But considerable problems remain. For example, it is by no means straightforward to determine the 'identity' of a complex, dynamical system which, by definition, is constantly changing. The criteria we use will depend on our subjective interest in the system, that is, on which aspects of system structure and function we assign greatest value to (Cumming and Collier 2005).

This example underlines the importance of continuously 'referring back' to the qualitative analysis that underpins key decisions taken in the process of designing the system model. The use of models to explore input-output relations in social-ecological systems must be undertaken as part of a wider, iterative epistemic cycle of exploration and reflection.

FRUITFUL HEURISTIC SYSTEMS CONCEPTS—POTENTIAL ENERGY LANDSCAPES

In order for models to be useful for policy-making and interdisciplinary dialogue it is helpful to be able to present the results of sophisticated mathematical analysis in simpler terms, without however losing sight of the underling concepts of dynamical systems. One heuristic which can be used for this purpose is the "potential energy landscape" that was proposed by Haken (1983), who developed "synergetics": If a ball in a landscape is taken as a symbol for the state of the system, the ball will roll downhill to the bottom of a valley (Figure 4.4A). If it is activated, it can propel to a hill in the landscape. In the language of systems theory, the valleys are "attractors" and the hills are "repellors". The term potential is equivalent to the potential energy. There is an "order parameter" ξ that drives the ball into various states. For instance, in a flat-bottomed valley (Figure 4.4B) the movement of the ball will fluctuate there for a while; this fluctuation can be the starting point for new trajectories. If the ball is located on hill between two valleys it is considered to be in unstable equilibrium, since only a small external force or random fluctuation is required to propel the ball into one or the other valley (Figure 4.4C). These various shapes of the landscape can be related to structural and behavioural properties of the system that are determined by a "control parameter" α that determines the shape of the landscape (Figure 4.4D).

Finally, in a qualitative description of the dynamics of the system valleys (or basins of attraction) can be characterised by their depth, the width and by their steepness; these characteristics determine the local dynamics of the system (Figure 4.4E). This concept is fruitful for discussing the concept of resilience (Holling and Gunderson 2002; Walker et al. 2004).

For instance, to take a purely hypothetical example, canopy height and biomass of a regenerating patch of forest could be the properties of a biosystem. Then the tendency of the ball to roll downhill represents the tendency of biomass to increase over time. The fork at the top of the slope represents a bifurcation point: a moment in the dynamic growth of the forest where

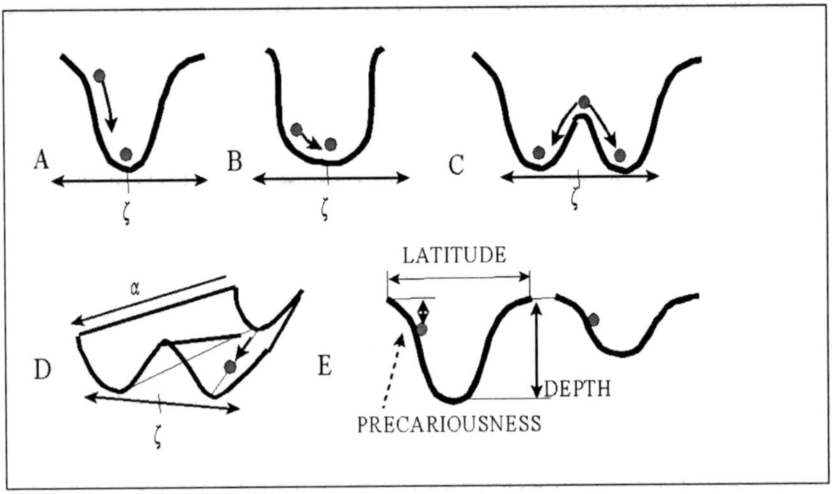

Figure 4.4 Potential landscape—a metaphorical scheme of complex system dynamics. The valleys (attractors) and hills (repellors) represent the potential of a ball to move from one point to another. The local shape of the landscape is determined by structural and dynamic properties of the system: If the parameter ζ varies, the ball will be activated but always rolls back to the bottom of the valley (A). If a second parameter ζ is varied the shape of the valley will change (B, C, D). The shape of the valley can be determined by variables such as latitude, depth and precariousness that jointly characterise the non-linear dynamics of the behaviour of the system.

two paths are possible: one leads to a low canopy and the other to a high canopy forest. Specific conditions at this point in time, such as the weather and seed availability, will determine which path is taken.

Such metaphorical models may provide a readily appreciable representation of key aspects of complex systems dynamics and are particularly useful for interdisciplinary communication between researchers who are less familiar with formal systems analysis.

Dealing with Unpredictability and Uncertainty

The principal function of input-output models is to identify the response characteristics of the system, that is, to answer 'how' and 'what if' questions. That is, we can specify one or more inputs into the system and determine probable outputs on the basis of theory and/or experience and build this relation into the model in the form of a mathematical function that can be expressed by an equation y=f(x). Nevertheless, the chain of causality leading from input to output cannot be known completely, due to the complexity of the processes involved, that is, the number of feedback loops. To simplify the analysis, and make it manageable, we treat the internal workings of the system as a black box (see Figure 4.5).

The same complexity limits the reliability of our model's predictions. In Heinz von Foerster's terminology, social-ecological systems are non-trivial (von Foerster 2002). Trivial systems are predictable, history independent and deterministic; non-trivial systems are history dependent and to a certain

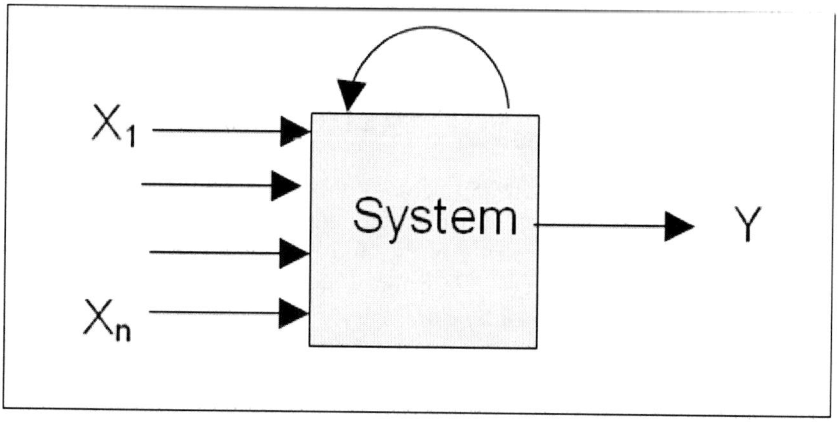

Figure 4.5 Black box analysis—the output of a system is related to multiple inputs. This results in a mathematical function that can be expressed by an equation y=f(x). The self-referring loop indicating a self-organising system complicates the analysis: The system behaves as a non-trivial machine with a 'transcomputational' number of states.

extent indeterminable and unpredictable. This non-trivial behaviour means that complex systems such as social-ecological systems have a 'transcomputational' number of states, so that precise calculations of system behaviour are not even theoretically possible (von Foerster 2002). Furthermore, it must always be remembered that in defining system boundaries, in order to make them amenable to formal systems analysis, we are 'bracketing out' a key feature of social-ecological systems in the real world: the un-bounded and ever-changing connectedness that exists within, between and across different scales. These inherent limitations in the accuracy of our models mean that although we can calibrate them to reproduce past patterns of behaviour, this does not assure that our predictions of future system behaviour will be correct (Haag and Kaupenjohann 2001). This doubt can be acknowledged, but only partially overcome, by expressing predictions in terms of probabilities of different events occurring within a wider range of possible outcomes of system behaviour.

How one responds to this uncertainty will depend to a large extent on the purpose for which the model has been constructed. If the purpose of the model is to provide inputs to policy and decision-making, its predictions can still be of value, to inform decisions made through adaptive management processes and the application of the precautionary principle (Haag and Kaupenjohann 2001). For academic work, the foregoing considerations do not negate the value of equation-based models for developing transdisciplinary understanding of the functioning and dynamics of social-ecological systems, from the perspective of self-organising complex systems. In systems science, formal modelling plays an analogous role to that of experimentation in traditional natural sciences, and is just as essential for the advancement of scientific understanding. The following section considers some principal methodological features of this approach towards learning about the world.

BASIC CONCEPTS FOR MODELLING SOCIAL-ECOLOGICAL SYSTEMS

As an epistemic object for human ecology (see Becker, this volume), the term 'social-ecological system' must be defined in a way which is compatible with the concepts, methods, theories and paradigms of the respective scientific approach. The precise definition will vary with the various conceptions and theories proposed by different authors and schools.

In principle, social-ecological systems can be characterised by relations between humans and nature. This agrees with definitions proposed by the Vienna School of Human Ecology (Knötig 1976, 1979) and by the German Society of Human Ecology[1] (Glaeser 1989; Glaeser and Teherani-Krönner 1992; Teherani-Krönner 1992; Steiner 1993), as well as by scholars associated with the Stockholm-based Resilience Alliance[2] (Anderies et al. 2004). An interesting variant proposed by the Frankfurt School of Social Ecology defines social-ecological systems as 'societal relations to nature' (see Becker

2009, this volume). This approach considers social-ecological systems as patterns of relations, rather than distinct 'entities'. Other variants are possible, for example, by substituting the terms 'society' for 'humans', or 'the environment' for 'nature'. But in any case, the starting point is generally the distinction between nature and human society. As Becker (this volume) notes, "without such a distinction the interaction between them is impossible."

Nevertheless, all the concepts in this basic definition are problematic. Nature and society are defined differently by the various scientific disciplines (e.g., by botanists, zoologists, agricultural economists, sociologists; comp. Weichhart 1993, 2008) and these will in turn differ from the various ways in which the terms are understood by lay practitioners (e.g., farmers, foresters or politicians). In considering nature, we are not dealing with nature per se but with a culturally determined perception and conception of nature. Transculturally, we seem to have an intuitive concept of nature as an autonomous system that surrounds us and is able to reproduce itself. But questions remain about the validity of 'nature' as a theoretical concept; for example, it can also be seen as an ideological subterfuge to justify the fiction that humanity occupies a special, privileged position on the planet, for instance, by Latour (1999, 2004, 2008) from a radical anthropological perspective and Gray (2002) from a radical Darwinist one. Even if one accepts the existence of nature and society as distinct features/subsystems of the world, it can be argued that there is a degree of interpenetration between them that makes the distinction difficult or impossible to apply for analytical purposes: "Nature is social [and] societies are natural" (Norgaard 1994: 36). The New Vienna School of Social Ecology has conceptualised the accelerating process of transformation of nature by human society as the 'colonization' of nature by society (Fischer-Kowalski 2004); the scale and impact of this transformation has led to the suggestion that we have entered a new, human-dominated geological epoch, the Anthropocene (Crutzen 2002). It would be possible to argue from here that there are only 'social systems' now; at the very least these ideas call into question the traditional assumption of the natural sciences that nature can be studied objectively, from the outside (see Becker 2009, and this volume).

Turning to the other substantive term in our definition, if we talk about the relation of humans to nature, we are implicitly considering human populations, and restricting our analysis to the material components of ecosystems (e.g., the impact of human populations on stocks of natural resources, or the effects on non-human species of pollution and landscape transformation). Few human ecologists would feel comfortable with this restriction; most would also be unwilling to consider human populations in this way as an undifferentiated mass. Even though by talking about 'humans' we have removed the gender bias contained in earlier discourse about 'man' and nature, we are still abstracting from other social divisions, such as social class, which are of course the central preoccupation of many sociologists. From a practical standpoint, focusing on populations makes it difficult to incorporate key issues such as social justice into our study of social-ecological systems.

The modern study of social systems does not focus on humans themselves, but on society as a text or a system of communications that provides the environment for human existence (Luhmann 1984; Luhmann et al. 1996). From this perspective, one could follow the Vienna School of Social Ecology in reformulating the definition of social-ecological system to focus on the relation between men (i.e., humans), nature and culture. What this definition implies is a differentiation of the term 'environment' into distinct natural (i.e., bio-geophysical) and cultural environments. Clearly, there is no reason to stop here, and further differentiations are possible into social, political, and technological environments, and so on. But following this path would imply, once again, progressively downgrading the autonomous role of nature in our system definition. From a modelling perspective, such a multipolar perspective would give rise to heterogeneous model complexity as explained in the following section.

Finally, the term 'relations' also needs to be specified. In its study of ecosystem dynamics, traditional ecology considers only material relations between system components, that is exchanges of matter and energy, and it is of course possible to study social-ecological systems, considered as relations between humans and nature, in the same way. However, most scholars now acknowledge the need to also consider non-material relations, which can be conceived of as exchanges of information (Janssen et al. 2006) or more generally as exchanges of meaning (Luhmann 1984; Luhmann et al. 1996; Halliday and Glaser 2011). This introduces an additional layer of complexity, but one that cannot be avoided if social aspects of the system are to be given due weight. These non-material relations can be described using sociological conceptions of interpersonal relations, such as roles, institutions, rules, laws, and so on (see Hummel, this volume), but these are often difficult to incorporate into the kinds of formal models which are the focus of this chapter.

The purpose of the preceding discussion is not to suggest that there are right or wrong definitions of nature and society, but to underline the fact that, in setting out to analyze social-ecological systems from a systems research perspective, the meanings of concepts like nature and society have to be defined precisely at the start (Tretter 1989; Serbser 2008). Systems thinking needs clear concepts and definitions of system boundaries, elements and their relations (Jax 2007); otherwise the resulting models will have only low heuristic value.

EXAMPLES OF QUALITATIVE MODELS OF SOCIAL-ECOLOGICAL SYSTEMS

Building models means the construction of representations of 'real-world' systems. Usually, the system components and the relations between them are described with the aim of explaining observed phenomena or predicting future ones. The simplest systems model consists of only two components; for instance one could model the connectivity of 'nature' and 'society'. Such a systems model contains at least two unidirectional relations linking the two components. Functionally speaking, the relations can have activating or inhibiting effects. A simple graphical model of a social-ecological system based on work by the Frankfurt School of Social Ecology focuses on the management of natural resources (Figure 4.6a). In this model, nature is seen by society as a resource for society but also for nature itself.

FIGURES 4.6a–c DIAGRAMS OF CONCEPTUAL FRAMEWORKS FOR UNDERSTANDING SOCIAL-ECOLOGICAL SYSTEMS

It is possible to differentiate the social-ecological system further into three basic components. Figure 4.6b represents the framework of the Vienna School and depicts a social-ecological system as relations between 'humans', 'nature' and 'culture'. These relations are societal metabolic processes—nature is experienced by men and is the object of work, and it is represented in culture that also provides a program for human action (Figure 4.6b). In this model we find eight unidirectional relations.

If four components are identified—for example, nature, population, society, technology—the model becomes more complex: 12 unidirectional relations can be identified. Such a conceptual framework was developed, for instance, by Duncan (Duncan and Schnore 1959; Duncan 1966) and gives rise to a model consisting of four components population, organisation, environment, and technology (POET). This framework is similar to a conceptualization of social ecosystems proposed by the first author of this chapter in the 1970s (Tretter and Schmid 1979; see Figure 4.6c).

As the number of components increases, the complexity rapidly reaches a level where computer simulations are required to understand the properties of the model. Global models have been developed that contain dozens or even hundreds of component variables (e.g., Forrester 1971; Meadows 1972; Meadows et al. 1992, Boumans et al. 2002).

The differences between the diagrams in Figure 4.6 show clearly how initial definitions of key concepts will influence the configuration of the resulting system model. Meta-analyses of models in social ecology are very rare. One approach was developed by Bernhard Glaeser (Glaeser 1994); more recently Glaser (2006) has undertaken a comparative review of different 'mind maps' of human nature relations. Glaeser distinguishes three basic types of models, namely, sociocentric, biocentric and integrative models (Glaeser 1994; see also Becker, this volume): Biocentric models focus on 'nature', which is differentiated into subsystems such as plants, animals, water, soil, air, and so on, and present 'humans' as a component within the

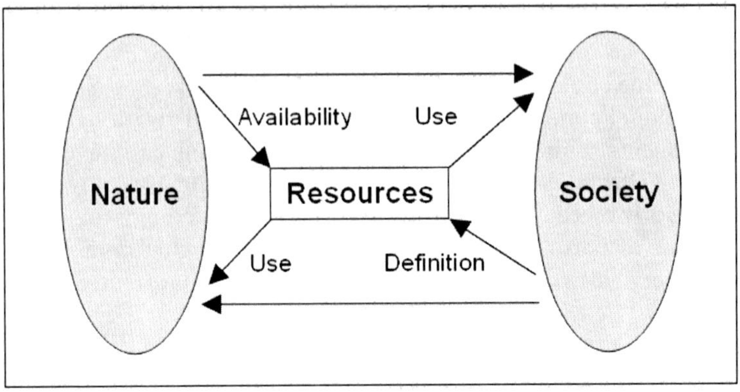

Figure 4.6a Model with two basic components and six interrelations.

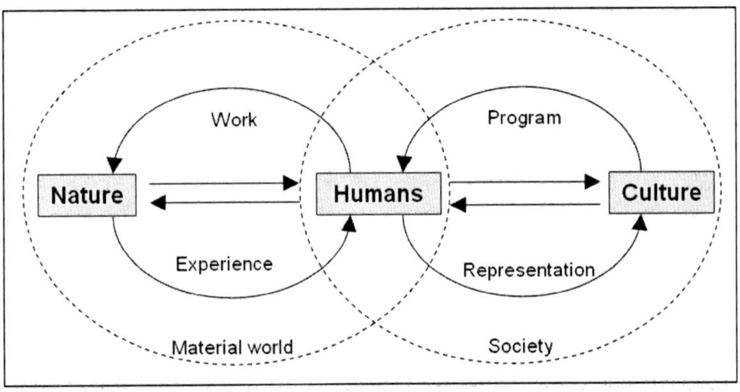

Figure 4.6b Model with three basic components and 8 unidirectional relations (Fischer-Kowalski 2004: 315).

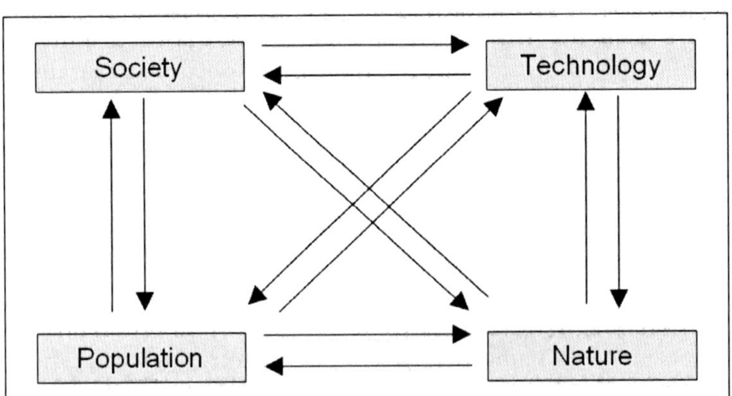

Figure 4.6c Frameworks with four basic components imply at least 12 unidirectional relations.

natural system, whereas sociocentric models focus on the social system (or society). Integrative models tend to represent a high conceptual resolution for both social and natural realms. In this regard, Glaser makes a further distinction between 'interdisciplinary' and 'complex system' mind maps (Glaser 2006).

The principal advantage of biocentric and sociocentric models is that the complications involved in multidisciplinary work can be minimised. Humans are conceptualised as part of nature, or nature is conceptualised as part of society, and the entire model can be analysed using methods from natural science or social science respectively. A fully integrated model will aim to integrate variables and data from sociology, economics, political science, geography, biology and other related disciplines. Therefore, theoretical 'bridge laws' are necessary in order to relate these heterogeneous variables to each other. At the same time, there is a need to proceed from the stage of qualitative modelling, as shown in the diagrams in this section, towards the formulation of explicit quantitative models.

The development of techniques for integrated, quantitative modelling of social-ecological systems should be a key aim of human ecology; however it is one that presents scholars with considerable theoretical, methodological and practical challenges. Some of these were discussed in the previous sections of this chapter. Now, we briefly switch to systems methodology.

DEVELOPING FORMAL MODELS FOR SYSTEMS SCIENCE

'Systems thinking' can be described as an approach to the study of situations characterised by organised complexity (László 1972). It has given rise to a large number of 'systems sciences' that apply a systems approach to specific realms of investigation in the natural and social sciences. 'Systems science' itself is the theory or science of systems (László 1972) and can be conceived of as a special interdisciplinary field of modelling and theory building (Ossimitz 2000; Imboden and Koch 2003; Bossel 2004; Tretter 2005; Fishwick 2007a, 2007b). It focuses on the multidisciplinary application of systems thinking and relies on specific concepts, methods, theories and paradigms (Tretter 2005). The foundational paradigms of systems science, which apply to all classes of systems, are hierarchy (Simon 1962), communication and control (Wiener 1948), and emergence (see Checkland 1981). From these foundations a number of different theoretical paradigms have been developed to explain or interpret the behaviour of complex systems, such as catastrophe theory, chaos theory, and complexity theory (Tretter 2005; Mainzer 1998; Érdi 2008).

It is important to mention in this context that this paradigm-centred approach to the advance of systems theory is in danger of losing sight of important basic ideas. Today 'complexity theory' is in fashion; some years ago 'non-linearity' and 'chaos' were the focuses of debate and, before this, 'catastrophe theory'. We have come a long way since the origins of systems theory, when 'control theory' was the main issue within the systemic-cybernetic paradigm. It is important to reaffirm the importance of control theory, and other foundational work in systems science, for the contemporary study of living systems, including social-ecological systems. An integrative view seems to be appropriate that considers systems thinking as a scientific discipline undertaken by a wide variety of specialists, whose work, however, is linked by the use of the same basic concepts (element, relation, non-equilibrium, complexity, etc.), methods (e.g., computer simulations), theories (control theory, chaos theory, complexity theory, etc.) and paradigms (self-organisation, cooperative phenomena, etc.)

The theories of systems science can be applied to the development of mathematical or qualitative 'meta-models' which provide theoretical frameworks for investigating the behaviour of particular classes of systems. The Lotka-Volterra model, which describes general features of the dynamics of interactive populations, is a famous example of a mathematical meta-model which has subsequently influenced systems thinking in ecology (Murray 2003, 2004). In sociology, an influential meta-model of this kind was the AGIL scheme of Talcott Parsons (Parsons 1951, 1971; Münch 1991, 1995, 1998). This conceptualises social systems in terms of the realization of four functions: adaptation (A), goal attainment (G), integration (I) and latent pattern maintenance (L). The 'adaptive cycle' is a widely known qualitative meta-model that informs much current thinking on social-ecological systems (Cumming and Collier 2005; Gunderson and Holling 2002).

Meta-models like these conceptualise general features of aspects of the world, abstracted from any particular situation or set of circumstances. They can be applied—more or less easily—for the development of a 'model', which in this context can be defined as a representation of specific, named features of the real world. A model attempts to describe structural relations, functional properties and other essential features of a real system of interest (Stachowiak 1973), normally with the aim of exploring, explaining or predicting the behaviour of the system over time.

Modelling today relies increasingly on formally defined methodologies (Ossimitz 2000; Sterman 2000; Hannon and Ruth 2001; Bossel 2004; Zeigler et al. 2007; Fishwick 2007a, 2007b). However in the context of social-ecological systems, it is notable that progress in developing formal models of the two realms has been uneven. In biological ecology, modelling has a long history and there are already established rules and principles (Odum 1994, 2007). If humans are included in ecological models, this makes possible the development of sophisticated 'earth system' models (Forrester 1971; Meadows 1972; Boumans et al. 2002). However these models usually 'bracket out' key sociological concerns, such as governance and equity. In social sciences (e.g., sociology), formal modelling is not common (but see Gilbert and Troitzsch 1999; Troitzsch and Gilbert 2006). Instead, theoretical reasoning is emphasised, and while this may lead to the formulation of 'meta-models', in the sense described previously, these are often difficult to apply in the development of formal models using established techniques. For example, Parson's AGIL scheme cannot be modelled easily and appears too abstract for use in empirical research (Parsons 1951, 1971; Münch 1995, 1998). One of the most influential theories of social systems is that of Niklas Luhmann (1984; Luhmann et al. 1996), but there are only a few examples of attempts to build computer-based models incorporating concepts from his system (Kron 2002). However, recent advances in computerised agent-based modelling open up the possibility of exploring emergent properties in social systems, by simulating the interaction of rule-based individual behaviour (see Sawyer 2005). This new methodological approach could of course be extended to include non-human agents for the computer-based exploration of human–nature interactions (Glaser, this volume).

MODELLING PROCEDURES

Building models is based on a stepwise procedure, shown schematically in Figure 4.7, although, as the diagram indicates, there is often a considerable amount of iteration between stages involved (Zeigler et al. 2007; Bossel 2004; Sterman 2000). The ultimate goal is a mathematically formulated model.

Systemic modelling starts with the identification of the systems that should be modelled. This involves an initial formulation of the system model at a conceptual level that is expressed by any form of mental map, verbal or graphical. The advantage of graphical representations, very often in the form of qualitative models consisting of boxes and arrows, is that they facilitate interdisciplinary communication. The system boundaries and system components must be defined, as well as the relations between them that

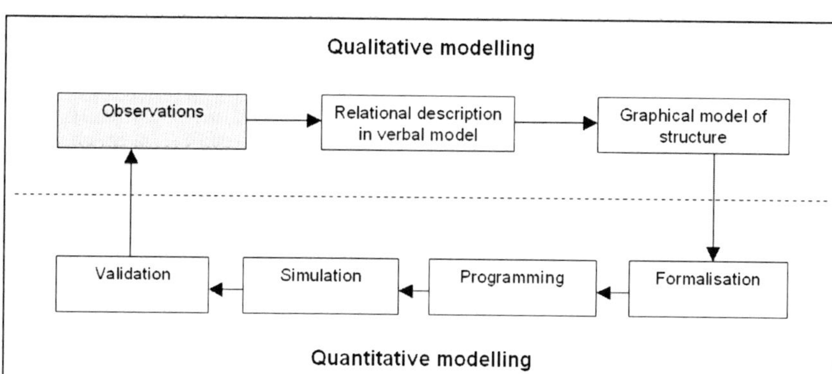

Figure 4.7 Steps in the development of a mathematical system mode.

are to be considered in the model. The initial conceptualization of the system configuration can be shown in system dynamics diagrams, such as the causal loop and stock and flow diagrams shown in Figures 4.8A and 4.8B. These indicate connections between elements of the model and the causal relations that either activate or inhibit their action. More advanced graphical languages have also been developed, such as Energy Systems Language, which can be used to represent more complex social-ecological processes (Montague 2007). In the next step, the model is formalised by translating these graphical conceptualizations into the language of mathematics: the relationships between the variables shown in the diagram are described in the form of mathematical equations. A further translation is required to transform the equations into a computerised model. Then quantified data is required to assign starting values for the components in the model. If data is unavailable, values can be assigned by expert-based estimation. In simulation, the model is run using past values as base-line data and the results for subsequent years are compared with the real, measured values for these years. This enables the model to be calibrated by modifying the equations on which it is based so as to reproduce past patterns of system behaviour. At this stage, future scenarios can be explored by running the model to simulate system behaviour under different hypothetical conditions, for example, representing different policy options. This will suggest possible answers to 'what if' questions that can be presented for discussion.

The global models of Jay Forrester and his co-workers were developed by such a stepwise procedures (Forrester 1971; Meadows 1972; Meadows et al. 1992); more recently, the development of a 'global unified metamodel of the biosphere' (GUMBO) has been described in detail by Boumans et al. (2002).

Computer-Based Modelling Applications

As noted already, one crucial step of the modelling procedure is the introduction of mathematics into the conceptual model. This stage can often present difficulties for the non-specialist. This section briefly reviews two computer-based approaches that can be applied by non-specialists using commercially available programs: icon-based modelling software and agent-based modelling.

In principle, any model of a complex real system can be composed of modules of activating and inhibitory components. Taking two elements of the system as a functional module, three possible types of modules can be distinguished: (1) an activator-inhibitor module, (2) an activator-activator module and (3) an inhibitor-inhibitor module. The first of these represents a negative feedback process and is characteristic of all living systems. The second type represents a positive feedback loop, and the third type exerts a behaviour that results in a suppression of the activity of one component combined with escalation of activity of the other component; in both cases leading to an unstable situation of unchecked growth or decline.

The relation between two variables can be expressed in the form of a simple differential equation. However, in real-world systems, each variable will be connected to may others and the equations expressing the relations between them become more complex. Many researchers in biological and social sciences have neither the specialist mathematical knowledge, nor the programming skills needed to transform the equations into 'code-based' (i.e., text-based) computer models. For this reason, special 'icon-based' modelling software has been developed since the 1960s, where models are constructed by assembling and linking model components (represented by various icons) by means of a sophisticated graphical user interface (GUI). The first of these was 'Dynamo', developed by Forrester. More recent examples include Stella™ (High Performance Systems Inc), POWERSIM™ (Powersim), Vensim™ (Ventana Systems Inc) and Model Maker™ (Cherwell Scientific Ltd). (See Sterman 2000 for details of these and other similar packages.)

This method of 'exploratory modelling' is briefly described here by referring to a partial model of the relation of population and food resources (Hannon and Ruth 2001):

- First, the conceptual model is represented in a causal loop diagram, in this case showing a negative feedback loop between population and food availability (Figure 4.8a). In other words, the more food is available, the more the population can grow; however, as the population grows, the availability of food decreases.
- A computer-based graphical model is then generated that incorporates our hypotheses about the behaviour of the system. For example in Figure 4.8b, the population is shown as being influenced by the number of births, while the number of births is influenced by population size both directly (more people means more births) and indirectly through the influence of population on the birth rate (the hypothesis here could be that a rising population is associated with a falling birth rate).
- The computer-based model is made up of quantifiable variables that are amenable to computer simulation. In Figure 4.8b these are population size, per capita food consumption (eat per capita), food production, birth rates, death rates, and so on. Values are then assigned to the variables drawing on empirical data and expert estimations.
- The model can then be run to simulate the behaviour of the system using different starting variables (Boccara 2004). This example shows a population explosion in an area with an abundance of food. But having consumed the available food the population is faced with starvation and therefore numbers decline. (Figure 4.8c).

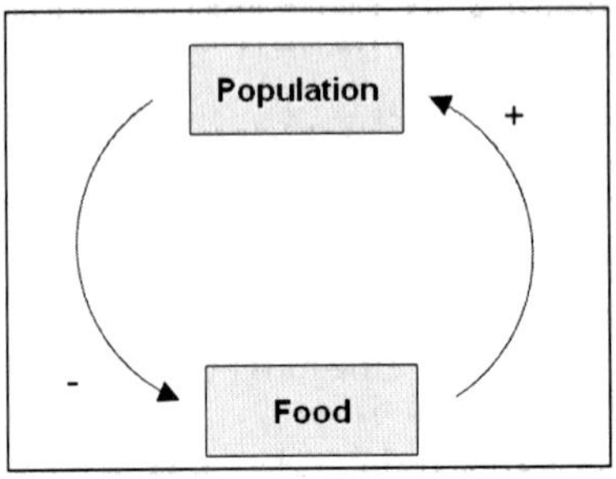

Figure 4.8a Causal loop diagram.

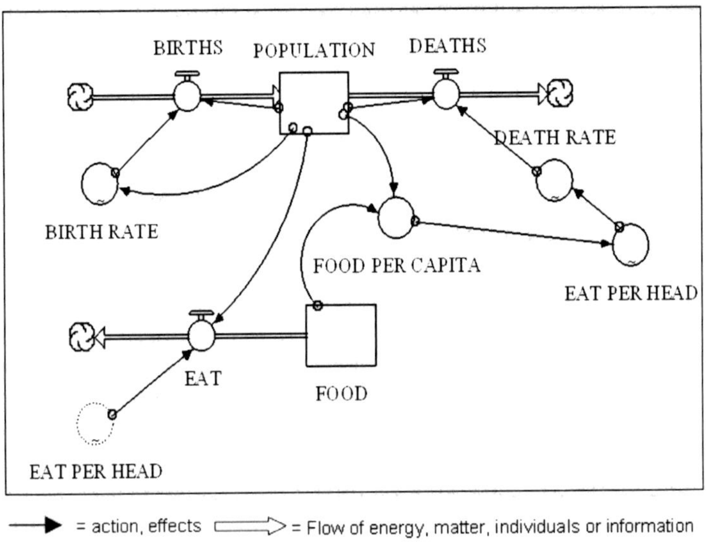

Figure 4.8b Stock and flow diagram (by Stella®).

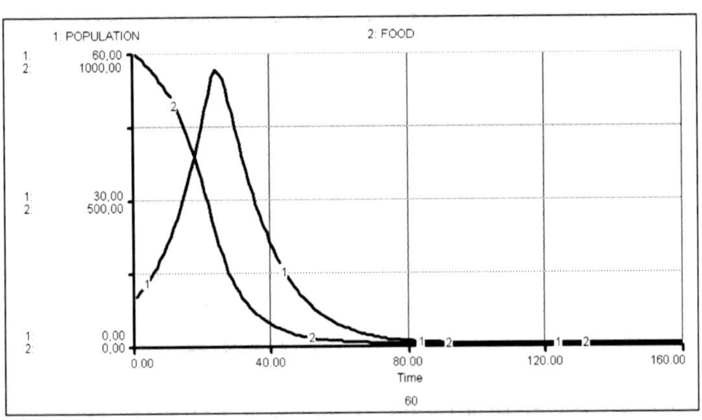

Figure 4.8c Modelling and simulation (by Stella®).

FIGURES 4.8a–c EXAMPLE FOR STEPS OF MODELLING: FROM QUALITATIVE CAUSAL LOOP DIAGRAM TO QUANTITATIVE SIMULATION

Usually, 'flight simulators' are used as computer interfaces where the user can change parameters, variables and coefficients in order to test the model by comparing its behaviour with the known behaviour of the real system that it represents. At this stage the necessary modifications can be made to calibrate the model. Then further changes can be introduced to explore how the system changes over time under different predicted or hypothetical conditions. For instance, the introduction of industrialised agriculture would change the relation between population and food availability and allow a higher population to persist over a longer period of time. At the same time, any negative effects envisaged as arising from this technological development could be incorporated into the model, and so on.

As noted already, the mathematical basis of such models of dynamic systems are differential equations or difference equations (Hannon and Ruth 2002). These can be used in a very wide variety of models, ranging from simple simulations of predator–prey interactions to the complex global models referred to previously. However, differential equations do have some limitations as an approach to modelling the behaviour of living systems. Firstly, in the real world change processes are discrete—not the continuous processes modelled by differential equations. Secondly, these models cannot show spatial dispersion phenomena. Furthermore, the properties of living systems are, to a greater or lesser extent determined by goal-oriented behaviour of individual agents—too few in numbers to assume a homogeneous population but too many to represent them easily in single differential equations.

For these situations, a second class of simulation programmes has been developed that enable dynamic spatial distributions to be modelled.

These models are known as agent-based models (Uhrmacher and Röhl 2007) and simulate the behaviour of cellular automata (Gardner 1970; Sloot and Hoekstra 2007). Once again a number of programs are commercially available of which NetLogo® is one of the most widely used (Wilensky 2001, 2009). The components of the models are agents that are programmed to operate autonomously in cellular (lattice-like) space according to rules determined by the modeller. For instance, the agent can be programmed to stop, or go or eliminate another agent in response to conditions in neighbouring cells. A famous prototype model is called 'Daisy World', where the development of the spatial distribution of populations of white and black daisies is simulated, by assigning simple rules for the birth and death of daisies under different conditions (Ford 2007). This method has succeeded in producing realistic reproductions of pattern formation in space and time by flocks of birds, ants and humans in traffic jams (Janssen and Carpenter 1999; Ratter, this volume). Figure 4.9a shows an example of the use of an

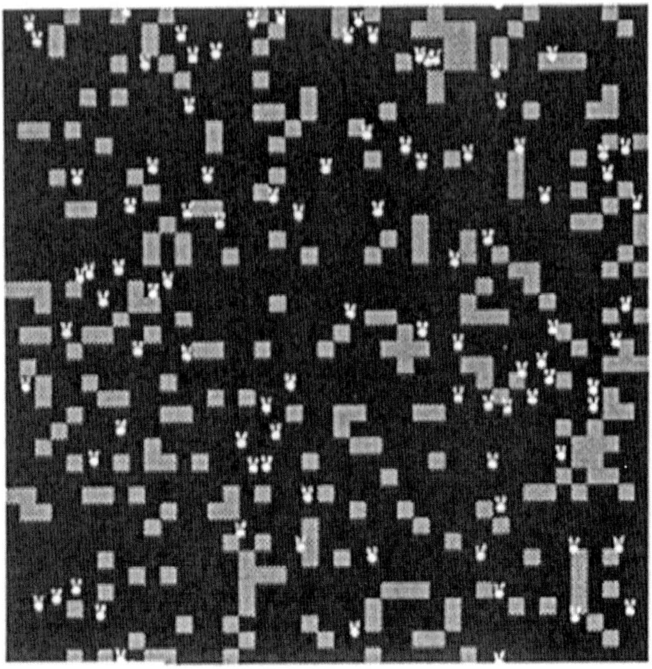

Figure 4.9a Screenshot of spatial distribution of rabbits and grass. NetLogo® Simulation program (Copyright 1999–2007 by Uri Wilensky).

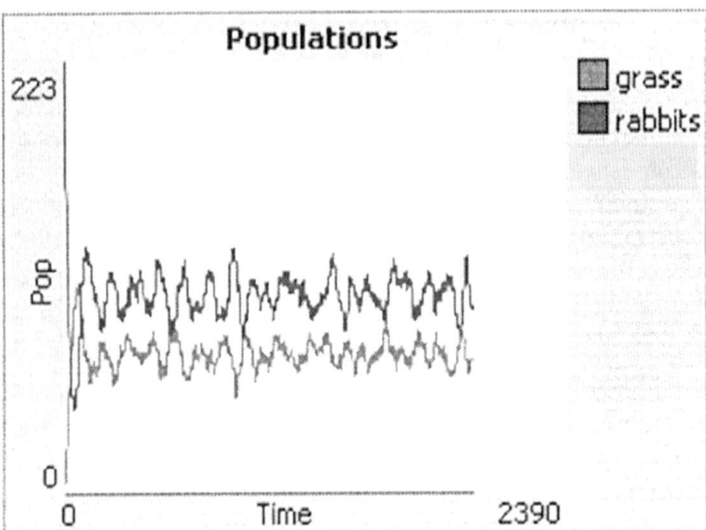

Figure 4.9b Time series of number of rabbits and amount of grass. NetLogo® Simulation program (Copyright 1999–2007 by Uri Wilensky).

agent-based model to produce a dynamic simulation of the spatial distribution of rabbits and grass. Standard icon-based modelling would have produced only the time-series information shown in Figure 4.9b.

Agent-based modelling has an obvious appeal to the social sciences since it offers the potential of modelling individual human decision-making that is not possible using traditional mathematical methods (Kron 2002; Sawyer 2005). Clearly, the methodology also has potential for modelling social-ecological systems. An advantage is that it explicitly considers human perception of and response to ecological changes, one of the challenges for social-ecological system modellers listed at the start of this chapter. Different options have been explored for ground-truthing and the iterative development of agent-based models with the participation of those whose actions are being modelled (Castella et al. 2005; Janssen and Ostrom 2006; Glaser, this volume; Wildenberg and Singh, this volume).

PERSPECTIVES

In a field such as human ecology where the scope for scientific experimentation is extremely limited, modelling acquires a central importance for the development and validation of scientific theory. This chapter has reviewed the challenges involved in moving from the verbal model, which still predominates in many social sciences, through qualitative conceptual models to formal mathematical and computer-based modelling. We have emphasised the function and value of models of all kinds as exploratory heuristic tools, and stressed the importance of starting out from a clearly defined conceptual framework in order to maximise the model's heuristic potential. Commercially available computer-based modelling programmes can play an important role in facilitating the exploration of the dynamics of social-ecological systems by multidisciplinary teams. However, the review of the modelling process highlights the need for rigour and clear thinking about the assumptions embodied in our models, both during the construction of the models and in the interpretation of the results.

NOTES

1. www.dg-humanoekologie.de
2. www.resalliance.org

REFERENCES

Anderies, JM, Janssen, MA, Ostrom. E (2004) A framework to analyze the robustness of social-ecological systems from an institutional perspective. Ecology and Society 9: Art. 18 (online)

Becker E, Jahn, T (2006) *Soziale Ökologie.* Campus, Frankfurt

Bertalanffy, L von (1968) *General System Theory*. Braziller, New York
Boccara, N (2004) *Modelling Complex Systems*. Springer, New York
Bossel, H (2004) *Systeme, Dynamik, Simulation*. Books On Demand, Norderstedt
Boumans, R, Costanza, R, Farley, J, Wilson, M, Portela, R, Rotmans, J, Villa, F, Grasso, M (2002) Modeling the dynamics of the integrated earth system and the value of global ecosystem services using the GUMBO model. Ecological Economics 41 (online)
Box GEP (1979) Robustness in the strategy of scientific model building. In: Launer, RL, Wilkinson, GN (eds) *Robustness in Statistics*. Academic Press, New York, p 202
Box, GEP, Jenkins, GM, Reinsel, GC (2008) *Time Series Analysis: Forecasting and Control*. Wiley, New York
Bühl, WL (1990) *Sozialer Wandel im Ungleichgewicht*. Enke, Stuttgart.
Bunge, M (1998) *Philosophy of Science*. Transaction Publishers, London
Campbell, DE (2004) Global transition to sustainable development. In: Ortega, E, Ulgiati, S (eds) Proceedings of IV Biennial International Workshop "Advances in Energy Studies. June 16–19, 2004". Unicamp, Campinas, SP, Brazil, pp 11–28
Cash, DW, Adger, N, Berkes, F, Garden, P, Lebel, L, Olsson, P, Pritchard, L, Young, O (2006) Scale and cross-scale dynamics: governance and information in a multilevel world. Ecology and Society, 11: Art. 8 (online)
Castella, JC, Tran Ngoc Trung, Boissau, S (2005) Participatory simulation of land-use changes in the northern mountains of Vietnam: The combined use of an agent-based model, a role-playing game, and a geographic information system. Ecology and Society 10: Art. 27 (online)
Checkland, Peter (1984) *Systems Thinking, Systems Practice*. Wiley, Chichester
Crutzen, PJ (2002) Geology of mankind. Nature 415: 23
Cumming, G, Collier, J (2005) Change and identity in complex systems. Ecology and Society 10: Art. 29 (online)
Deaton, ML, Winebrake, JI (2000) *Dynamic Modeling of Environmental Systems*. Springer, New York
Duncan, OD (1966) Human ecology and population studies. In: Hauser, PM, Duncan, OD (eds) *The Study of Population*. University Chicago Press, Chicago, pp 678–716
Duncan, OD, Schnore, LF (1959) Cultural, behavioral and ecological perspectives in the study of social organization. American Journal of Sociology 65: 132–146
Érdi, P (2008) *Complexity Explained*. Springer, New York
Fischer-Kowalski, M (2004) Gesellschaftliche Kolonisierung natürlicher Systeme. Arbeiten an einem Theorieversuch. In: Serbser, W (ed) *Humanökologie: Ursprünge—Trends—Zukünfte*. Oekom Verlag, München, pp 308–325
Fishwick, PA (2007a) The languages of dynamic system modelling. In: Fishwick, PA (ed) *Handbook of Dynamic System Modelling*. Chatman & Hull, New York, pp 1.1–1.12
Fishwick, PA (2007b) *Handbook of Dynamic System Modelling*. Chapman & Hull, New York
Foerster, H von (2002) *Understanding Systems: Conversations on Epistemology and Ethics*. Springer, Berlin
Folke, C, Pritchard, L Jr, Berkes, F, Colding, J, Svedin, U (2007) The problem of fit between ecosystems and institutions: Ten years later. Ecology and Society 12: Art. 30 (online)
Ford, AF (2007) System dynamic models of environmental systems. In: Fishwick, PA *Handbook of Dynamic System Modelling*. Chapman & Hull, New York, pp 27.1–27.17
Forrester, JW (1971) *World Dynamics*. MIT Press, Cambridge, MA

Gardner, M (1970) The fantastic combinations of John Conway's new solitaire game "life". Scientific American 23: 21–321

Giddens, A (1976) *New Rules of the Sociological Method: A Positive Critique of Interpretative Sociologies*. Hutchinson, London.

Gilbert, N, Troitzsch, KG (1999) *Simulation for the Social Scientist: Concepts in the Social Sciences*. Open University Press, Berkshire

Glaeser, B (1989) *Humanökologie*. Westdeutscher Verlag, Opladen

Glaeser, B (1994) *Environment, Development, Agriculture: Integrated Policy Through Human Ecology*. University College London Press, London

Glaeser, B, Teherani-Krönner, P (1992) *Humanökologie und Kulturökologie*. Westdeutscher Verlag, Opladen

Glaser, M (2006) The social dimension in ecosystem management: Strengths and weaknesses of human–nature mind maps. Human Ecology Review 13: 122–142

Gray, JN (2002) *Straw Dogs: Thoughts on Humans and Other Animals*. Granta Books, London

Haag, D, Kaupenjohann, M (2001) Parameters, prediction, post-normal science and the precautionary principle—a roadmap for modelling for decision-making. Ecological Modelling 144: 45–60

Haken, H (1983) *Synergetics, an Introduction: Nonequilibrium Phase Transitions and Self- Organization in Physics, Chemistry and Biology*. Springer, New York

Halliday, A, Glaser, M (2011) A management perspective on social ecological systems: A generic system model and its application to a case study from Peru. Human Ecology Review 18: 1–18

Halliday, MAK (2005) On matter and meaning: The two realms of human experience. Linguistics and the Human Sciences 1: 59–82

Hannon, B, Ruth, M (2001) *Dynamic modelling*. Springer, Berlin

Hill, DRC, Coquillard, P (2007) Ecological modeling and simulation: From historical development to individual based modeling. In: Fishwick, PA (ed) *Handbook of Dynamic System Modelling*. Chapman & Hull, New York, pp 29.1–29.18

Holland, JH (1998) *Emergence: From Chaos to Order*. Helix Books (Addison-Wesley), Reading

Holling, CS, Gunderson, LH (2002) Resilience and adaptive cycles. In: Gunderson, LH, Holling, CS (eds) *Panarchy: Understanding Transformations in Human and Natural Systems*, Island Press, Washington, DC, pp 25–62

Imboden, DM, Koch, S (2003) *Systemanalyse*. Springer, Berlin

Janssen, MA, Bodin, Ö, Anderies, JM, Elmqvist, T, Ernstson, H, McAllister, RRJ, Olsson, P, Ryan, P (2006) A network perspective on the resilience of social-ecological systems. Ecology and Society, 11: Art. 15 (online)

Janssen, MA, Carpenter, SR (1999) Managing the resilience of lakes: A multi-agent modelling approach. Conservation Ecology 3: Art. 15 (online)

Janssen, MA, Ostrom, E (2006) Empirically based, agent-based models. Ecology and Society 11: Art. 37 (online)

Jax, K (2007) Can we define ecosystems? On the confusion between definition and description of ecological concepts. Acta biotheoretica 55: 341–355

Korzybski, A (1994) Science and Sanity: An Introduction to Non-Aristotelian Systems and General Semantics. Institute of General Semantics, Fort Worth, TX

Kron, T (2002) *Luhmann modelliert. Sozionische Ansätze zur Simulation von Kommunikationssystemen*. Leske und Budrich, Opladen

László, E (1972) *Introduction to Systems Philosophy: Toward A New Paradigm of Contemporary Thought*. Harper Torchbooks, New York

Latour, B (1999) *Die Hoffnung der Pandora*. Suhrkamp, Frankfurt

Latour, B (2004) *Politics of Nature: How to Bring the Sciences into Democracy*. Harvard University Press, Cambridge, MA, London

Latour, B (2008) *Wir sind nie modern gewesen*. Suhrkamp, Frankfurt
Liehr, S (2004) Integrated water quality management—development of a social-ecological approach. Vortrag gehalten bei dem DFG-Workshop on Integrated Water Research and Water Management, 28/29.6.2004
Liu, J, Dietz, T, Carpenter, SR, Alberti, M, Folke, C, Moran, E, Pell, AC, Deadman, P, Kratz, T, Lubchenco, J, Ostrom, E, Ouyang, Z, Provencher, W, Redman, CL, Schneider, SH, Taylor, WW (2007) Complexity of coupled human and natural systems. Science 317: 1513–1516
Luhmann, N (1984) *Soziale Systeme*. Suhrkamp, Frankfurt
Luhmann, N, Bednarz, J, Baecker, D (1996) *Social Systems*. Stanford University Press, Stanford, CA
Mainzer, K (1998) *Thinking in Complexity*. Springer, Berlin
Meadows, DL (1972) *Limits to Growth*. Universe Books, New York
Meadows, DH, Randers, J, Meadows, DL (1992) *Limits to Growth: The 30-Year Update*. Chelsea Green Publishing, White River
Midgley, G (2000) *Systemic Intervention: Philosophy, Methodology and Practice*. Kluwer Academic / Plenum Publications, New York, Boston, Dordrecht, London, Moscow
Montague, CL (2007) Dynamic simulation with energy system language. In: Fishwick, PA (ed) *Handbook of Dynamic System Modelling*. Chapman & Hull, New York, pp 2.8.1–2.8.32
Münch, R (1991) *Dialektik der Kommunikationsgesellschaft*. Suhrkamp, Frankfurt am Main
Münch, R (1995) *Dynamik der Kommunikationsgesellschaft*. Suhrkamp, Frankfurt am Main
Münch, R (1998) *Globale Dynamik lokale Lebenswelten*. Suhrkamp, Frankfurt am Main
Murray, JD (2003) *Mathematical Biology: Spatial Models and Biomedical Applications*. Springer, New York
Murray, JD (2004) *An Introduction to Mathematical Biology: An Introductory Course*. Springer, New York
Norgaard, R (1994) *Development Betrayed: The End of Progress and a Coevolutionary Revisioning of the Future*. Routledge, London, New York
Odum, HT (1994) *Ecological and General Systems: An Introduction to Systems Ecology* (Rev. ed.). University Press of Colorado, Denver
Odum, HT (2007) *Environment, Power, and Society for the Twenty-First Century: The Hierarchy of Energy*. Columbia University Press, New York
Ossimitz, G (2000) *Die Entwicklung systemischen Denkens*. Profil, München
Parsons, T (1951) *The Social System*. Free Press, Glencoe
Parsons, T (1971) *The System of Modern Societies*. Prentice Hall, Engelwood Cliffs
Richmond, B (2001) *An Introduction to Systems Thinking*. High Performance, Hanover
Sawyer, RK (2005) *Social Emergence: Societies as Complex Systems*. Cambridge University Press, New York
Serbser, W (2008) Die Natur der Gesellschaft. In: Bruckmeier, K, Serbser, W (ed) *Ethik und Umweltpolitik*. Oekom-Verlag, München, pp 51–74
Simon, HA (1962) The architecture of complexity. Proceedings of the American Philosophical Society 106: 467–482
Sloot, PMA, Hoekstra, AG (2007) Modeling dynamic systems with cellular automata. In: Fishwick, PA (ed) *Handbook of Dynamic System Modelling*. Chapman & Hull, New York, pp 21.1–21.19
Stachowiak, H (1973) *Allgemeine Modelltheorie*. Springer, New York

Steiner, D (1993) Human ecology as trasnsdisciplinary science. In: Steiner, D, Nausser, M (eds) *Human Ecology—Fragments of Antifragmentary Views of the World*. Routledge, London, pp 47–76

Sterman, J (2000) *Business Dynamics*. McGrawHill, New York

Sterman, J (2002) All models are wrong: Reflections on becoming a systems scientist. Systems Dynamics Review 18: 401–531

Teherani-Krönner, P (1992) Von der Humanökologie der Chicagoer Schule zur Kulturökologie. In: Glaeser, B, Teherani-Krönner, P (ed) *Humanökologie und Kulturökologie*. Westdeutscher Verlag, Opladen, pp 16–43

Tretter, F (1979) Cybernetics—a revised conception and its relevance for theoretical human ecology. In: Knötig, H (ed) *Proceedings of the Second International Meeting on Human Ecology, Vienna 1977*. St. Saphorin, Georgi, pp 209–232

Tretter, F (1982) On the development and multidisciplinary relevance of a qualitative analytical systems technology for biology, psychology and sociology. In: Trappl, R, Klir, J, Pichler, F (eds) *Progress in Cybernetics and Systems Research, Vol. 3*. McGraw Hill, New York, pp 179–184

Tretter, F (1989) Grundprobleme des Begriffs "Umwelt". Natur- und Ganzheitsmedizin 7: 193–201

Tretter F (1998) *Ökologie der Sucht*. Hogrefe, Göttingen

Tretter F (2005) *Systemtheorie im klinischen Kontext*. Pabst, Lengerich

Tretter, F, Schmid, J (1979) On the development of a new conception of social ecology. In: Knötig, H (ed) *Proceedings of the Second International Meeting on Human Ecology, Vienna 1977*. St. Saphorin, Georgi, pp 233–242

Troitzsch, KG, Gilbert, N (2006) *Systems Dynamics and World Models*, ZUMA Simulation Workshop 09.2006. Koblenz

Uhrmacher, A, Röhl, M (2007) Agent-oriented modelling in simulation: agents for modelling, and modelling for agents. In: Fishwick, PA (ed) *Handbook of Dynamic System Modelling*. Chapman & Hull, New York, pp 8.1–8.16

Walker, BH, Holling, CS, Carpenter, SC, Kinzig, AP (2004) Resilience, adaptability and transformability. Ecology and Society 9: Art. 5 (online)

Watson, AJ, Lovelock, JE (1983) Biological homeostasis of the global environment—the parable of daisy world. Tellus 35B: 284–298

Weichhart, P (1993) How does the person fit into the humanecological triangle? From dualism to duality—transactional view. In: Steiner, D, Nausser, M (eds) *Human Ecology—Fragments of Antifragmentary Views of the World*. Routledge, London, pp 77–104

Weichhart, P (2008) *Entwicklungslinien der Sozialgeographie*. Franz Steiner, Stuttgart

Wiener, N (1948) *Cybernetics*. MIT Press, Cambridge, MA

Wilensky, U (2001) *Modeling Nature's Emergent Patterns with Multi-agent Languages*. (online: http://ccl.northwestern.edu/uri/public_html/papers/MEE, 10.05.04)

Wright, GH v. (2004) *Explanation and Understanding*. Cornell University Press, Ithaca, NY

Zeigler, BP, Praehofer, H, Kim, TG (2007) *Theory of Modelling and Simulation*. Elsevier India private limited, New Delhi

5 Complexity and Emergence
Key Concepts in Non-Linear Dynamic Systems

Beate M.W. Ratter

INTRODUCTION

Complexity and emergence are buzzwords in our modern world. But what do they mean? And how can they help us to see and to understand the world in a different manner? Complexity and emergence are key concepts in the study of non-linear dynamic systems. Theories about these 'complex systems' attempt to explain how relationships between components of the system give rise to patterns of collective behaviour that define the properties of the 'whole system', which interacts and forms relationships with its environment. A complex system contains many autonomous components which interact with each other—for example, human society and the economy, insect colonies, cellular automata on computers, some chemical systems, and other systems in various domains. Nature and society are two examples of dynamic, non-linear systems. Both realms can be defined in terms of the individual elements they are composed of and by the relations between these elements. Many of these elements will themselves be complex systems; the large-scale systems are composed of a hierarchy of subsystems.

In all these systems, global behaviour emerges from the local interactions between the participants. This behaviour can be studied with reference to structural and functional properties of whole system, based on the hypothesis that these are more relevant than specific details about the behaviour of individual system components. By contrast, the study of complexity tries to explain the emergence of macro-scale phenomena as the product of non-linear and dynamic interactions between micro-scale elements (Mainzer 2003, 2007). "Emergence is the reason why there are hurricanes, and eco-systems, and complex organisms like humankind, not to mention traffic congestion and rock concerts. Indeed, the term is positively awe-inspiring" (Corning 2002: 2). But what is emergence? What does it really explain? And why is it so readily welcomed, in spite of its opacity, by many different scientists, whether system thinkers or not?

Scientific use of the terms complexity and emergence needs to be based on formal definitions and a consideration of their limitations in specific research contexts. It is important to distinguish between a situation where complexity itself is the research object, that is, we are engaged in the 'study of systems', and one where the aim is to use the concept of complexity in the study of traditional scientific disciplines, such as biology or sociology, to gain insights and new perspectives on our objects of study in these fields (Egner, Ratter and Dikau 2008). In the latter case, the findings of complexity research need to be translated to facilitate their application in these new contexts. This chapter is a contribution to this ongoing work.

A promising field for the application of the findings of complexity theories is the analysis of human–nature interactions. Natural and social systems are dynamic and non-linear. In considering the interaction of both systems we are dealing with coupled social-ecological systems. In order to analyse such systems we have to acknowledge the interdependency that exists between subsystems and elements on different scales with different spatial and temporal characteristics (Berkes et al. 2003; Egner and Ratter 2008) and the fact that interaction between the elements on one scale can lead to emergent phenomena on another scale. The theory of complexity, which is rooted in chaos theory, provides a new perspective on the behaviour of such systems as well as the interaction between different systems. This concept encourages a *Gestaltwechsel*[1]—a change in perception—which helps to see things differently and fosters the search for new answers to emerging questions about the interaction of human–nature interface.

WHAT IS COMPLEXITY?

The term 'complex' is used in many different ways and encompasses a great variety of phenomena. In daily life the word 'complex' is mainly used when we mean 'complicated'. But complex is not the same as complicated. Complex is not "*even more complicated than complicated*" (Ratter 2006). The terms complexity, complex and complex systems are currently used widely in the scientific community. The multifarious use and the lack of a distinct and unambiguous definition pose a risk that these terms will become empty phrases. Drawing on the mathematical definition of Kolmo-gorov complexity, Okubo adopts a quantitative approach to complexity and states: "The more complex a model becomes, the more parameters are involved" (Okubo 1980: 6). In contrast to this understanding of complexity, there are definitions which focus on the kinds of relations between system elements. Manson (2001) considers what kinds of interaction between elements of a system can give rise to complex behaviour. This behaviour is shaped by the relations between the entities which make up the system, the processes occurring within the system, its internal structure and self-organisation of the system to adapt to conditions in the external environment. Berkhoff, Karstens and Newig (2004) describe complexity as a state between order and chaos with a high degree of structural diversity. They explain that characteristic properties of complex systems are the result of the co-existence of many heterogeneous entities, the dynamics of the system, the process of spontaneous self-organisation and emergent behaviour (see also Mainzer 2003).

The different examples make clear that there are at least two fundamentally different ways to define complexity. On the one hand, emphasis is placed on the number of elements of a system, and on the other, the focus is on the relationships and the processes occurring between the elements of the system. In other words, one approach focused on the constitution of the system and the other on its behaviour.

STRUCTURAL COMPLEXITY VERSUS BEHAVIOURAL COMPLEXITY

The computer scientist Schamanek (Folke) makes the distinction between structural and behavioural complexity. When a system consists of a lot of different elements and interactions, this is structural complexity. Behavioural complexity arises from processes and relations between the system elements. However, behaviourally complex systems are not necessarily composed of large numbers elements and relations between them. That is, the complex behaviour of a system does not automatically imply a complicated structure. A system that exhibits behavioural complexity can consist of simple *or* complicated structures (Ratter 2001). This point can be illustrated by considering the example of the behavioural complexity of a double pendulum. A double pendulum is a very simple physical system. It consists of a pendulum A attached to another pendulum B. Each pendulum is a simple mechanical system. However, joined together, the double pendulum AB exhibits complex behaviour (see Selinger 2000 and Figure 5.1).

This is an example of the application of a theory of complexity which refers to the *behaviour*, not only to the structure of a system. Simple systems can show a complex behaviour; complicated systems with many elements are structurally complex *and* can show a complex behaviour—then they are structurally *and* behaviourally complex. The fundamental question is, How can something complex arise from the simple? It is assumed that behaviour cannot be understood solely with reference to the properties of single elements but that it can be explained from the interactions of the system elements. Even simple systems can have a complicated behaviour, for example, the double pendulum. But structurally complex systems such as the weather or the stock exchange can show complex behaviour too, which is determined by the interaction of many single elements or agents as parts of the system. In such dynamic systems the elements (e.g., agents) interact in a non-linear way. Complicated counterintuitive behaviour can emerge through non-linear interactions. A complex system can display surprising properties when elementary sub-processes are coupled together to form a larger system (Casti 1994: 40).

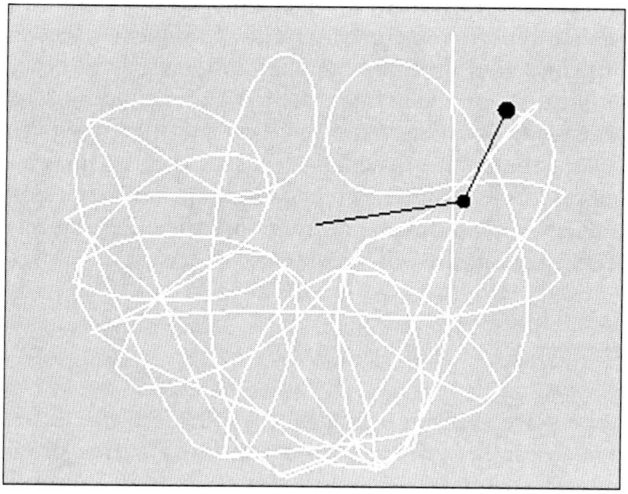

Figure 5.1 Double pendulum with canvas traces (Selinger 2000).

Behavioural complexity considers the qualities and properties of a system as well as the relations and processes between the elements. Fundamental in this is the recognition that non-linear behaviour of a complex system is irreducible. The behaviour of the system is to a high degree dependent on its history. It is unpredictable because of its non-linear and dynamic character. The behaviour of complex systems can be random (Mainzer 1997, 2003; Waldrop 1992). The attributes of self-organisation ensure the self-preservation of the system; however, the structure of the system can change over time through internal feedback processes and adaptation to changes in the system environment. In effect, the identity of the system can change over time. Therefore, knowledge about the history of the system is fundamental in order to understand its present character. A system is behaviourally complex if it possesses *key characteristics* and not only *key variables*. The ongoing processes of structure building in a complex system are unpredictable. Structural reorganisation has no single cause, but arises from micro-scale interdependencies that act together to cause major changes: the spontaneous appearance and disappearance of order and system structures. The emphasis is on the '*becoming*'. Complex means becoming, development, emergence—which can lead to profound qualitative transformations (Ratter 2006). This is the point where the theory of complexity becomes practical relevance in the analysis and planning of human–nature interactions. I shall come back to this issue later.

In order to recognise patterns in systems with emergent characteristics, it is necessary to shift to non-linear thinking; linear thinking will not help to understand the interactions between system elements and even less the resulting patterns of system behaviour because of the irreducible character and unpredictability of emergent systems. Complexity theory incorporates a set of principles that can be applied in the study of different types of complex system. It provides the basis for understanding *how* patterns are formed due to the local and repeated interaction between system components and *how* these patterns change. In this context, interactions and the consequences of these interactions are significant. It is not assumed that systems exist in equilibrium and therefore the changing relationships between system elements are of importance.

SURPRISE AS PART OF THE SYSTEM

Dynamic systems are systems in which a stable state can occur and disappear abruptly. Stability is represents a transitory period of equilibrium within the longer-term alternation of the system between pattern and non-pattern phases. In order to understand systems undergoing constant changes surprise should be accepted as an inherent feature of system dynamics (Ratter 2006: 116). Qualitative transformations can occur as a result of internal system processes and changing relations between the elements of the system, without external stimulus. Therefore, emergent characteristics cannot be understood without reference to relations between the components of the system (Manson 2001: 406).

"Complex systems involve the collective behaviour of a large group of relatively simple elements or agents, in which the interactions among these elements typically are local and non-linear" (Bedau and Humphreys 2008: 209). Based on the theory of complexity, complex systems show emergent macro-scale structures, which evolve from the interaction of the constituent parts and subsystems (Janssen 2002). Or as Mainzer (1997) puts it, referring to Aristotle, "[T]he whole is more than the sum of its parts" or "the whole is not only the sum of its parts". This basic theorem of system thinking emphasises the fact that identifying system elements is not sufficient in order to understand a complex system. The deconstruction of a system can provide insights but suppresses the interactions between the system elements (Ratter and Treiling 2008). In the analysis of coupled social-ecological systems this suppression could have serious consequences.

WHAT IS EMERGENCE?

Emergence is something which we do not expect but which we experience in different ways and situations. An illustrative example of the phenomenon 'emergence' in social systems is the phenomenon of traffic congestion. We all get hooked up in traffic jams once in a while—but do we ask ourselves why this happens? Traffic congestion can emerge without a specific cause such as a car accident, a broken bridge or road works. Even without these blocking events we get trapped in traffic jams, merely as a result of car drivers following simple rules. For example, drive a at certain speed and do not crash into the car in front; slow down if there is a car close ahead, speed up if not.

> Traffic jams can start from small *seeds*. These cars start with random positions and random speeds. If some cars are clustered together, they will move slowly, causing cars behind them to slow down, and a traffic jam forms. Even though all of the cars are moving forward, the traffic jams tend to move backwards. The behaviour of the group is often very different from the behaviour of the individuals that make up the group. (Wilensky 1997)

The concept of emergence can be traced back to an argument of Aristotle in his *Metaphysics*: "The whole is something over and above its parts, and not just the sum of them all" (see Aristotle reprint 1974). This implies an ontological distinction between parts and wholes. For Goldstein (1999: 51) emergence can be defined as "the arising of novel and coherent structures, patterns and properties during the process of self-organisation in complex systems" (see also Corning 2002). Emergence has been used by physicists to explain Bénard (convection) cells, by psychologists to explain consciousness, by economists to explain stock market behaviour, and by organisation theorists to explain informal 'networks' in large companies.

However, emergence can also be understood as the appearance of something unexpected. According to Corning (2002: 24) the overwhelming majority of people understand the term emergence as a synonym for "appearance" or "growth", rather than as a referring to the relation between parts and wholes. Example of emergence in this sense are democracy in Russia, soccer as a school sport in the US, the Internet, mad cow disease, and so on.

The two different concepts of emergence can lead to confusion in the definition and use of the term. Emergence can be understood as *appearance of something*, and also in the Aristotlean sense as *a phenomenon of synergy*—"a parts–whole relationship" (Corning 2002: 24). In complexity theory emergence is associated with dynamic systems whose behaviour arises from the interaction among its parts and cannot be predicted from knowledge about the parts in isolation (Casti 1997). In the context of a paradigm shift from stability to change, from equilibrium to non-equilibrium, and from continuity to discontinuity, processes in nature and social systems are perceived differently. Corning (2002) calls this phenomenon the "*re-emergence of emergence*". Hundreds of years after Aristotle, the perception of stability and equilibrium is finally being questioned again and the classical perception of continuity in natural development is being overcome. In consequence, the Darwinian dictum ought to be rephrased as *natura facit saltum*!

Corning himself considers that it is difficult to pinpoint the exact date of the re-emergence of emergence as a mainstream concept,

> but it roughly coincided with the growth of scientific interest in the phenomenon of complexity and the development of new, non-linear mathematical tools—particularly chaos theory and dynamical systems theory—which allowed scientists to model the interactions within complex, dynamic systems in new and insightful ways. (Corning 2002: 23)

Nowadays, emergence is considered in theories about various natural and social sciences.

Emergent structures are patterns not created by a single event or rule. Nothing commands the system to form a pattern. Instead, the interaction of each part with its immediate surroundings causes a complex chain of processes leading to some sort of order. One might conclude that emergent structures are more than the sum of their parts because the emergent order will not arise if the various parts are simply coexisting; the interaction between these parts is central. Emergent structures can be found in many natural phenomena, from the physical to the biological domain, for example, weather phenomena such as hurricanes are emergent structures (see Bedau and Humphreys 2008).

Fenzl (2003) considers emergence as the appearance of a new property of a system which can not be deduced or previously observed as a functional characteristic of the system. "For example, water has emergent properties different from its interconnected parts (molecules of H and O). These properties disappear if the molecules are separated again" (Fenzl 2003: 252). Self-organisation is the appearance of new system structures without explicit pressure from outside the system, that is, from the system environment.

STRONG VERSUS WEAK EMERGENCE

Emergence can be divided into two perspectives: weak emergence and strong emergence. Weak emergence describes new properties arising in systems as a result of the interactions at an elemental level. Emergence, in this case, is merely part of the model that is needed to describe a system's behaviour. According to Stephan (1999) "weak emergentism" is characterised by three postulates: physical monism, systemic properties and synchronous determinism. These postulates mean that there are no supernatural powers or entities which decide on how the system should behave, that systems may have properties that none of their parts have, and that these properties are coupled to its micro-structure. This means that the properties of a system only change as a result of a change in its structure or in the properties of its parts. Brunner and Klauninger (2003: 28) argue that weak emergentism is practically indistinguishable from reductionism.

On the other hand, systems can have qualities not directly traceable to the system's components, but rather to how those components interact. These new qualities are irreducible to the system's constituent parts and their properties (Laughlin 2005). This view of emergence is called strong emergence. Regarding strong emergence, Bedau (1997) observes,

> Although strong emergence is logically possible, it is uncomfortably like magic. How does an irreducible but supervenient downward causal power arise, since by definition it can not be due to the aggregation of the micro-level potentialities? Such causal powers would be quite unlike anything within our scientific ken. This not only indicates how they will discomfort reasonable forms of materialism. Their mysteriousness will only heighten the traditional worry that emergence entails illegitimately getting something from nothing.[2]

From this point of view emergence is a phenomenon which evolves from the interaction of constituent elements of a system, but which cannot be explained only from the single elements and its properties. One calls emergence counterintuitive because emergent structures are generated by processes of self-organisation inside a system without external influences.

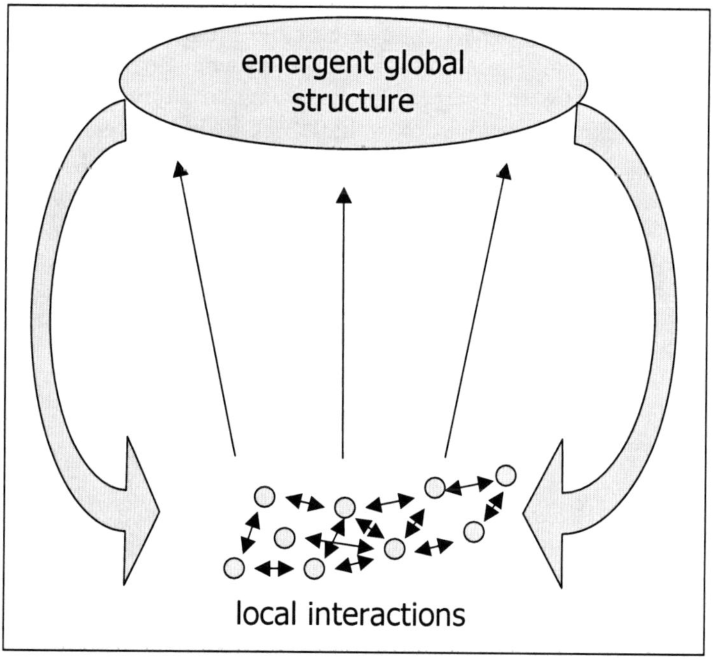

Figure 5.2 Abstract illustration of emergence in complex systems (accord. Lewin 1993: 25).

The non-linear interactions between the composing parts of the system give rise to a high-level, global emergent structure or organisation that is not straightforward to explain from the local components (see Figure 5.2).

The occurring emergent behaviour impacts those components which evolved the emergent behaviour due to a mechanism of feedback in the system. A number of systems share this kind of architecture. They contain many individuals/elements/agents, each one of which interacting with its neighbours. From the network of these local interactions, global features emerge (up-going arrows): market forces, cultural and social trends, features of insect colonies, and so on. The global features define an environment which influences the rules of interaction of the single participant with its neighbours (down-going fat arrows). To understand a complex system one has to analyse the higher level of emergent collective behaviour as well (Lewin 1993).

The problem when dealing with these kinds of systems is the impossibility of predicting how changes in the rules governing individual interactions will affect global system behaviour. For example, a new law that aims to change the conduct of individual participants in a social system may give rise to global side effects, which will finally yield a collective behaviour that is just the opposite of what was originally intended. Thus, to act effectively in such domains, it is important to understand the ways in which a global behaviour emerges from the collection of many local interactions. There are many open questions here, and one of the most important is: which exactly are the systems that exhibit this kind of behaviour?

Complex systems with emergent qualities have some coherent behaviour for a certain period of time (Goldstein 1999). This coherence correlates the level of the producing entities into a unity on the level of emergence. Emergent qualities are not pre-given, but the result of the dynamical development of complex systems. The notion of emergence means that a system is more than the sum of its parts and that a developing system has new qualities that cannot be reduced to old states or prior existing systems. A key cause of emergent phenomenon is therefore the interaction of elements or agents of a system.

NOT ACTORS, BUT AGENTS

Single, locally acting persons can be pioneers or initiators of macro-level developments—they act as *agents* in the sense of the theory of complexity. An agent is a unit or element of the system which acts locally and which cannot directly influence the system globally. Agents are restricted in their actions, which, however, have system-changing character through iteration and repetition. They create spatial clusters of quasi-symbiotic unities or combating complexes (Casti 1994; Ratter 2001). An agent is not the same as an actor. *Actor* is the superordinate concept. *Agent* is a special type of actor without or with incomplete knowledge about the system as a whole. Its activities can only affect areas beyond his local level indirectly, that is, via some form of mediation. It is not chance that the term *agent* is also sued to describe spies, in the sense of 'secret agent'. Agents or spies engaged in the cloak-and-dagger games of the former superpowers had similarly restricted knowledge, about only their own operational areas. Their local activities on the micro-level, however, could have global significance and influence the macro-level behaviour of the system (Ratter 2001: 57ff).

This type of system, composed by interacting agents, can be analysed formally under certain conditions (Casti 1995/1996: 7ff). One approach is by way of simulation. Simulation is possible (1) if the agents possess only limited local information and (2) if agents possess restricted information processing capacity and (Folke) if the system itself is composed of a limited number of interacting agents—enough to give rise to interesting structures; but few enough for simulation to be possible. The analysis of these types of systems can be approached by multi-agent simulation (MAS) (see Mandl 2000; Glaser and Krause, this volume). Simulation can be used to understand how a system has behaved in the past—for backcasting—and to evaluate proposed policies by forecasting their impact on system behaviour. A simulation is nothing more and nothing less than a model of systems.

These systems are significant because local interactions between system elements have a determining effect on the history of the system (Epstein and Axtell 1996; Kochugovindan and Vriend 1998). Nature and society can be conceived as the product of interactions between agents: as non-linear dynamic systems or the emergence from the non-linear interaction of agents. The focus on interactions fosters a *Gestaltwechsel*—a new perception of systems as coupled micro and macro behaviour that can help explain the emergent properties of systems.

WHY COMPLEXITY?

"Complex systems exhibit emergent properties in order to adapt to changing circumstances in an effective and efficient manner, they adaptively self-organise" (Bullock and Cliff 2004: 4). Complex systems show key attributes such as dynamics, non-linearity and emergence. These attributes should be part of the research objective since they are fundamental to an understanding of dynamic systems. System development is history dependent and contingent: the identification of bifurcation points, behavioural jumps and surprises is a key challenge in order to understand the behaviour of a dynamic system. In research into human–nature interactions we deal with non-linear systems and try to understand how they work. The trajectories of non-linear systems display discontinuities and cannot be predicted with any degree of confidence, as they are the result of strong emergence from the interaction of the constituent agents and elements. Bifurcation points are fundamental turning points where the system state is poised between two alternative future paths of development. Uncertainties and surprises are part of the system's history and not exceptions. And system's behaviour is heavily path-dependent. Each system has its own history.

An analysis aimed at only the structural complexity of a system cannot encompass the dynamic character of non-linear systems. The interaction of society and nature is therefore not sufficiently described and understood if one restricts it to direct causality. Behaviourally complex systems are characterised by intensive processuality. This means that 'reality' of such systems is constantly changing and they can not satisfactorily be described with reference to their invariant elements (Eisenhardt et al. 1995). The key characteristics of complex systems (dynamic processes, non-linearity, emergence) have to be considered as part of the system and accepted as part of the system's history and behaviour. The identification of the bifurcation points can help us to understand current system behaviour and contribute to the determination of policy measures that facilitate adaptation by the system to environmental changes (Ratter 2001; Treiling 2009).

WHY EMERGENCE?

Emergence and self-organisation are strong paradigms in the theory of complexity which draw attention to the new perspective of systems analysis. The increasing use of these concepts in natural as well as in social sciences and in manifold disciplines, ranging from brain research to economics, indicates that the classical reductionist approach in science is being replaced by a new paradigm. The question is no longer *how* to reduce complicated issues to simple processes but *how* can become something complex arise from the simple? The explanatory paradigm (A is just the same as B) changes towards the exploratory paradigm (can A create B?). These ideas have even found their way into politics which there is more an and interest in the possibility that certain services can be provided without hierarchy and central state influence by using the self-organising forces of the civil society. The new research objective is the translation of mechanisms of individual activity in divergent systems' trajectories and searching for the reasons of divergent planning cultures in different societies. At this point the theory of complexity acquires practical relevance (Ratter 2001: 55, 237).

The behaviour of agent-based non-linear systems results in uncertainty, which we experience in daily life and which we have to take into account especially in strategic planning. The system trajectory cannot be controlled. Lau (2006) declares, "Systems can not be controlled but tickled." One can try to set frames or give impulses in order to encourage the system to change. But one cannot presage what the system is going to do next, but trust that it will do something. The idea of preordaining the system's development should be abolished as well as the hope of understanding system behaviour completely. Nevertheless, if change is an inevitable result of the agents' interaction, one could at least attempt to learn from these processes (Lau 2006).

PLANNING IN SPITE OF UNCERTAINTY

Planning is the intention to construct certainty in an uncertain environment. In a complex and non-linear system planning security does not exist. The system behaves dynamically and its development is continuously non-linear. Planning takes places in the context of unpredictable system dynamics. It has to be adaptive and can only make short-term predictions about how the system will develop from its current state (Mainzer 1997, 2003). In this context, the most appropriate planning approach is an adaptive management, which accepts non-linear dynamics and emergence as a fact of life and adopts an iterative adaptation process based on monitoring and feedback. One step beyond adaptive management is *complex management*, which in addition takes account of the social and cultural background of a society, different perceptions and the political framework (Ratter 2001). Planning takes place in a socio-cultural environment with interacting agents tickling the system's trajectory in innumerable ways. Under these circumstances planning becomes the management of uncertainty—dealing with non-linear coupled human–nature systems and making preparations for changes and surprises.

SUMMARY

Theories of complexity can contribute to a *Gestaltwechsel*—they can help us to look at systems differently, to search for new questions and new answers. A new perspective in systems analysis facilitates the consideration and analysis of coupled natural and social systems, which are both dynamic and non-linear. In these systems it is the iterative activity of agents which has the greatest influence on the system's trajectory. Agents and bifurcation points are decisive for the history of a system. Process simulation on the basis of MAS can lead to a deeper understanding of system dynamics—but cannot change the unpredictable development of reality. Surprises have to be accepted as inherent to the system. Linear thinking can be dangerous in a non-linear world.

An aspect of non-linear dynamic systems are that big changes which occur are not the result of single big causes but more emerge out of micro-scale-interdependencies. There is spontaneous emergence and disappearance of order and structures. Theories of complexity provide a set of principles that can be applied to a range of complex systems. They enable us to describe *how* patterns are formed out of repeated local interactions between components and *how* these patterns change. The focus is on the description of behaviour, regardless of whether the make up of the system is simple or complicated. It is assumed that the behaviour of systems can be explained from the interactions taking place between the system elements. But the emphasis is on "becoming" and on qualitative discontinuities.

In order to examine emergence, complexity research focuses on non-linear relationships instead of focusing on linear relationships (e.g., of energy and information). "This stocks-and-flows perspective emphasises quantities of flow and not necessarily its quality. Complexity research employs techniques . . . to examine qualitative characteristics such as the symbolic content of communication" (Manson 2001: 406).

System behaviour is contingent. An emergent phenomenon can be defined as a 'subset' of the vast universe of cooperative interactions that produce synergistic effects of various kinds, both in nature and in human societies. "In this definition, emergence would be confined to those synergistic wholes that are composed of things of 'unlike kind'. It would also be limited to qualitative novelties—i.e., unique synergistic effects that are generated by functional complementarities, or a combination of labour" (Corning 2002: 28). In order to understand such systems where the system state is constantly changing, surprise must be accepted as part of the system dynamics (Ratter 2006: 116).

However, it should not be understood from the forgoing that theories of complexity are a justification for doing nothing or for not interfering. Rather they should be seen as underlining the importance of including a wide range of agents in the planning process. "The problem is that doing nothing does not necessarily stabilise the equilibrium of a complex system and can drive it to another metastable state" (Mainzer 1997: 324). The most important thing to know about complexity is simple: "There is still something to come" (Lotter 2006: 46). Emergence has to be included in our line of thoughts.

NOTES

1. The concept is based on the fundamental principle of *gestalt perception* according the Berlin School of Gestalt Psychology (see Bruce, Green and Georgeson 1996).
2. For further explanations and distinction of strong and weak emergence see Bedau (2008); see also Arshinov and Fuchs (2003).

REFERENCES

Arshinov, V, Fuchs, C (2003) *Causality, Emergence, Self-Organisation.* NIA-Priroda, Moscow

Aristotle (reprint 1974) *Metaphysik*. Rowohlt, Reinbek bei Hamburg

Bedau, MA (1997) Weak emergence. In: Tomberlin, J (ed) Philosophical Perspectives: Mind, Causation, and World, Vol. 11. Blackwell Publishers, Oxford, pp 375–399

Bedau, MA (2008) Downward causation and autonomy in weak emergence. In: Bedau, MA, Humphreys, P (eds) *Emergence: Contemporary Readings in Philosophy and Science*. Bradford Book, Cambridge, MA, pp 155–188

Berkes, F, Colding, J, Folke, C (2003) *Navigating Social-Ecological Systems*. Cambridge University Press, Cambridge

Berkhoff, K, Karstens, B, Newig, J (2004) Komplexität und komplexe adaptive Systeme—Ansätze des Santa Fe Institut. In: Bundesministerium für Bildung und Forschung (ed) *Steuerung und Transformation. Überblick über theoretische Konzepte in den Projekten der sozial-ökologischen Forschung* (Diskussionspapier 01 der Querschnittsarbeitsgruppe Steuerung und Transformation im Förderschwerpunkt sozial-ökologische Forschung des BMBF). Berlin, pp 101–107

Bruce, V, Green, P, Georgeson, M (1996) *Visual Perception: Physiology, Psychology and Ecology* (3rd ed.). Laurence Earlbaum Associates, Hove, UK

Brunner, K, Klauninger, B (2003) An integrative image of causality and emergence. In: Arshinov, V, Fuchs, C (eds) *Causality, Emergence, Self-Organisation*. NIA-Priroda, Moscow, pp 23–35

Bullock, S, Cliff, D (2004) Complexity and emergent behaviour in ICT systems. Technical Report HP-2004-187, Semantic & Adaptive Systems, Hewlett-Packard Labs. (Unpublished). (online: http://eprints.ecs.soton.ac.uk/11478/, 14.12.2008)

Casti, JL (1994) *Complexification: Explaining a Paradoxical World Through the Science of Surprise*. HarperPerennial edition, New York

Casti, JL (1995/1996) Seeing the light at El Farol. Complexity 1: 7–10

Casti, JL (1997) *Would-Be Worlds: How Simulation Is Changing the Frontiers of Science*. John Wiley, New York

Corning, PA (2002) The re-emergence of 'emergence': A venerable concept in search of a theory. Complexity 7: 18–30

Egner, H, Ratter, BMW (2008) Wozu Systemtheorie(n)? In: Egner, H, Ratter, BMW, Dikau, R (eds) *Umwelt als System—System als Umwelt*. oekom, München, pp 9–20

Egner, H, Ratter, BMW, Dikau, R (eds) (2008) *Umwelt als System—System als Umwelt*. oekom, München

Eisenhardt, P, Kurt, D, Stiehl, H (1995) *Wie Neues entsteht. Die Wissenschaft des Komplexen und Fraktalen*. Rowohlt, Hamburg

Epstein, JM, Axtell, R (1996) *Growing Artificial Societies: Social Science from the Bottom up*. Brookings/MIT Press, Washington, DC

Fenzl, N (2003) Emergence and self-organisation of complex systems. The role of energy flows and information. In: Arshinov, V, Fuchs, C (eds) *Causality, Emergence, Self-Organisation*. NIA-Priroda, Moscow, pp 245–258

Goldstein, J (1999) Emergence as a construct: History and issues. Emergence 1: 49–72

Janssen, M (2002) *Complexity and Ecosystem Management: The Theory and Practice of Multi-Agent Systems*. EE, Cheltenham

Kochugovindan, S, Vriend, NJ (1998) Is the study of complex adaptive systems going to solve the mystery of Adam Smith's "Invisible Hand"? The Independent Review—A Journal of Political Economy 3: 53–66

Lau, P (2006) Alles funktioniert. brand eins 8: 111–116

Laughlin, R (2005) *A Different Universe: Reinventing Physics from the bottom down*. New York, Basic Books

Lewin, R (1993) *Die Komplexitätstheorie—Wissenschaft nach der Chaosforschung*. Hoffmann und Campe, Hamburg

Lotter, W (2006) Einfach mehr. brand eins 8: 46–55

Mainzer, K (1997) *Thinking in Complexity*. Springer, Berlin, Heidelberg

Mainzer, K (2003) *Thinking in Complexity: The Computational Dynamics of Matter, Mind, and Mankind* (4th ed). Springer, Berlin, Heidelberg

Mainzer, K (2007) *Der kreative Zufall. Wie das Neue in die Welt kommt.* C.H. Beck, München

Mandl, P (2000) Geo-Simulation—Experimentieren und Problemlösen mit GIS-Modellen. In: Strobl, J, Blaschke, T, Griesebner, G (eds) *Angewandte Geographische Informationsverarbeitung XII.* Heidelberg, Herbert Wichman Verlag, pp 345–356

Manson, SM (2001) Simplifying complexity: a review of complexity theory. Geoforum 32: 405–414

Okubo, A (1980) *Diffusion and Ecological Problems: Mathematical models.* Springer, Berlin, Heidelberg

Ratter, BMW (2001) *Natur, Kultur und Komplexität: Adaptives Umweltmanagement am Niagara Escarpment Ontario, Kanada.* Springer, Berlin, Heidelberg

Ratter, BMW (2006) Komplexitätstheorie und Geographie—Ein Beitrag zur Begründung einer anderen Sicht auf Systeme. Mitteilungen der Österreichischen Geographischen Gesellschaft, 148. Jg.: 109–124

Ratter, BMW, Treiling, T (2008) Komplexität oder was bedeuten die Pfeile zwischen den Kästchen? In: Egner, H, Ratter, BMW, Dikau, R (eds) *Umwelt als System—System als Umwelt.* oekom, München, pp 23–37

Schamanek, A (1998) *Umwelt Management Austria: Einführung in Komplexe Dynamische Systeme.* (online: http://www.ams.smc.univie.ac.at/~schamane/kds/ov.htm, 25.03.2007)

Selinger, P (2000) Double pendulum. Department of Mathematics and Statistics. Dalhousie. (online: http://www.mathstat.dal.ca/~selinger/lagrange/doublependulum.html, 23.05.08; last update of page 12.05.2000).

Stephan, A (1999) *Emergenz—Von der Unvorhersagbarkeit zur Selbstorganisation.* Dresden University Press, Dresden

Treiling, T (2009) *Mensch/Natur-Interaktion im Oberen Mittelrheintal—Komplexitätstheoretische Ansätze am Beispiel des Weinbaus.* Dissertation am Geographischen Institut der Johannes Gutenberg-Universität, Mainz

Waldrop, MM (1992) *Complexity: The Emerging Science at the Edge of Order and Chaos.* Touchstone Simon & Schuster, New York

Wilensky, U (1997) *NetLogo Traffic Basic Model.* Center for Connected Learning and Computer-Based Modeling, Northwestern University, Evanston, IL. (online: http://ccl.northwestern.edu/netlogo/models/TrafficBasic, 30.04.08)

Part III

Resilience, Adaptability and Transformability

6 Archetypes of Adaptation to Climate Change

Klaus Eisenack

INTRODUCTION

The potential impacts of climate change are expected to be substantial but regionally diverse. While it is highly certain that, for example, precipitation patterns will somehow change, it is not as clear which regions will become wetter or dryer. Depending on place, such biophysical changes will meet different socio-economic and institutional conditions and therefore have different effects. While the political debate on appropriate reactions to this challenge has mainly focussed on mitigation of greenhouse gas emissions, climate change is already occurring and will persist for centuries. Therefore, adaptation to existing or potential changes is receiving increasing attention. The Intergovernmental Panel on Climate Change (IPCC) defines adaptation as "adjustment in natural or human systems in response to actual or expected climatic stimuli or their effects, which moderates harm or exploits beneficial opportunities" (IPCC 2007a: 869). It notes that "a wide array of adaptation options is available, but more extensive adaptation than is currently occurring is required to reduce vulnerability to future climate change. There are barriers, limits and costs, but these are not fully understood" (IPCC 2007b: 19). One example for such a barrier is the local diversity of impacts and vulnerabilities already mentioned. This makes it difficult for local decision-makers to recognise the need for action and to learn from the experience of actors in other places (e.g., Lecocq and Shalizi 2007; Eisenack 2010).

Standard economic analysis frames climate protection as a global public good, while adaptation is seen as a private good (e.g., Nordhaus 1990). Everyone profits from reduced damage due to slowed climate change, but the costs of greenhouse gas mitigation are borne only by those who take action to protect the climate. As a result of this tension, it is known that public goods are insufficiently provided for by private actors. Due to the global nature of the problem, it cannot be resolved by a single government acting alone. By contrast, it is claimed that the costs and benefits of adaptation both lie with those that adapt. These arguments imply that local, regional or business actors should prefer adaptation to mitigation activities. In this light, the

IPCC's reference to "barriers and limits" to adaptation is remarkable since it indicates that current trends are not in line with theoretical expectations. The emergence of various bottom-up climate change mitigation activities in recent years (e.g., Cities Alliance 2007) contributes to this puzzle. A further example is the "Schwarzenegger effect" (Urpelainen 2009), which refers to the Californian state government's adoption of international climate protection initiatives, bypassing the inactive federal government.

This chapter aims to contribute to solving this puzzle and to an understanding of barriers to adaptation that are not captured by the standard analysis. It starts out from the hypothesis that the diversity of local conditions is one crucial barrier to adaptation. This hinders the transfer of insights about adaptation from one case to the other. There seems to be no single recipe that offers a general solution to the problem of how to overcome barriers to adaptation. A pessimistic conclusion would be that situations have to be evaluated on a case-by-case basis, which would increase the costs of adaptation substantially. On the other hand, the impression remains that not every case of potential adaptation is utterly different, but that typical constellations of local circumstances, hindering adaptation, re-appear. But how should one go about dealing with locally diverse problems and find locally appropriate solutions, when common features can nevertheless be identified? Can methodological proposals to deal with these questions help us to understand and respond to local under-adaptation to ongoing climate change?

I address these questions by giving an overview of the archetype approach to interdisciplinary research about social-ecological problems. This approach sets out to identify reappearing functional patterns of change (archetypes) in a systematic way. Although, until recently, it has been applied in other, rather different contexts, it seems tailored for questions of the aforementioned type. The paper sketches out a preliminary set of archetypes relating to barriers to adaptation. To achieve greater precision in understanding what is meant by adaptation to climate change and in the description of the proposed archetypes, I introduce the action theory of adaptation as a conceptual basis.

THE ARCHETYPE APPROACH

Local citizens and decision-makers are nowadays confronted with broad information on climate change, which is nevertheless difficult to comprehend, prioritise and process with the limited resources or capabilities they have available. If stakeholders have a basic awareness of the problem, and the opportunities and threats it brings, they often ask whether they should react at all, and if so, what to do. However, blueprint solutions are often not available, and successful practices depend to a high degree on the local context, that is, location-specific biophysical and socio-economic dynamics.

In principle, the context-dependence of social-ecological dynamics makes it difficult to draw general conclusions about determinants of effective governance. This is the prevailing view (e.g., Warren 2002; Schachhuber 2004), including studies of theoretical ecology (e.g., Shrader-Frechette 1993). Institutional design principles meant to be generic can only achieve this objective if they are formulated in a very abstract way (Schachhuber 2004) or over-simplify the contextual conditions (Ostrom 2007). This means that considerable refinement is required when they are applied to concrete problems, raising the question of whether they are of any practical use, beyond theory. On the other hand, promising results have been reported which do not provide the basis for a grand theory, but can be seen as representing a collection of generic factors that favour success (e.g., Ostrom 1990). It is thus possible to compare cases of successful and unsuccessful governance. To be useful for establishing best practices and transferring experience, such a comparison should not be too coarse-grained. This requires a notion of similarity at an appropriate level of abstraction.

A related challenge is the integration of knowledge from quantitative modelling and qualitative case study research. Quantitative analyses are often criticised for over-generalizing and it is claimed that they disregard context by hiding processes behind data and variables. In contrast, qualitative studies are criticised for being vague and of limited use since the results are not transferable to other cases. There appears to be a dichotomy between quantitative (statistical, mathematical, computational) and qualitative (interpretative, dialectic, hermeneutic) methods. Typically, quantitative work is identified with nomothetic, and qualitative with idiographic research (cf. Windelband 1919).

Archetypes are representative patterns of the interaction between society and nature that bring about environmental change and/or are brought about by such changes. They illustrate basic underlying processes and help to make connections between what is happening in different regions and to help decision-makers to understand their particular situation within a broader context. They are building blocks of social-ecological interaction that reappear in multiple case studies, meaning that they can be found at different places around the world because these places share certain conditions. Starting from the premise that every place is particular, speaking of reappearing patterns requires that the patterns are described in terms which are sufficiently abstract to cover relevant properties of multiple specific cases. This abstraction should be general enough to be potentially found in more than one case, but not so abstract that it applies to every case (which would make it meaningless). One criterion for identifying the appropriate degree of generalization is derived from the underlying assumption that successful institutional arrangements can be transferred between cases if they share archetypes. This approach is fundamentally different from deriving a grand theory that covers all cases, and from idiographic descriptions of single cases.

The usage of the term 'archetype' is similar to the use of the word 'pattern', although both terms are more crisply defined in different scientific domains (e.g., Kelso 1997). Its meaning refers to complex objects of cognition, for example, as "ideas of modes and relations" (Locke 1690), but also to the common notion of a "primordial image, character, or pattern of circumstances that recurs throughout literature and thought consistently enough to be considered universal" (Encyclopædia Britannica 2006). In this chapter, the term does not refer to the psychology of Jung (1954), although he probably introduced the term for similar reasons. Speaking of an 'archetype' instead of a 'type' is justified since the term refers not to a mere collection of entities that have common properties, but to a complex of relations that reappears in parts of multiple systems, even of different types. It is a description of a functional pattern that merits attention for being invariant over long time scales. Alternatively, we can use the term to denote a defining example for a certain type, and subsume different instances by similarity to this example. In this sense, the notion of an archetype resembles that of an ideal type (Weber 1922), and the reappearance of an archetype is not recognised by its belonging to an abstract equivalence class as indicated above, but by family resemblance (Wittgenstein 1953) to a paradigmatic case.

Archetypes are described as building blocks because it is not required that every case can be completely described by a single archetype. That does not mean that single cases can be explained by different alternative archetypes, but that they cover social-ecological systems only partially. While a single case can be considered as a functional unit of society–nature interactions and relations, each archetype covers only a selective part of these relations. For a comprehensive picture of a single case it is in general necessary to combine several archetypes. This generalization criterion, which refers to shared archetypes, only makes sense in this context. When two cases can partially be described by the same pattern, this does not imply that they appear completely identical from this abstracted perspective.

For a more detailed presentation and discussion of the archetype approach, see Eisenack et al. (2006). However, the basic ingredients of archetype analysis are not totally new. The aspiration to distil general knowledge by comparative analysis is at least as old as modern science. The particular feature of archetype analysis is that it aims to understand phenomena, in this case the dynamics of social-ecological systems, in a way that is as general as possible, but fine-grained enough to account for local particularities.

In systems dynamics, paradigmatic models of management problems were developed under the term of archetypes (e.g., Wolstenholme 2003). These models are formulated using causal-loop diagrams and aim at deriving generic solutions. These systems archetypes lead some authors to make interesting distinctions between different ways of conceiving generic structure (Lane and Smart 1996). Ragin (1987) employs Boolean analysis of qualitative data to a selection of abstract variables for the study of political science and history. His qualitative case study analysis (QCA) derives a general logical formula for the analysis of all dependent variables of a selected set of cases. It can be transformed into an equivalent statement in so-called disjunctive normal form, that is expressed as follows: "[T]he property described by the dependent variable holds for all cases where . . . or . . . or . . . ". Here a list of alternative conditions is linked by the logical 'or'. This resembles the statement that not a single archetype can explain every case. Lambin et al. (2003) used a similar approach for extensive meta-studies to explain processes of desertification and tropical deforestation. This work is also motivated by the desire to preserve the descriptive richness of local case studies, while contributing to a general understanding of the issue. Their comparative method identifies configurations of causes with similar outcomes. Ostrom (1990) and Ostrom (2007) develop a set of explanatory variables and institutional design principles that contribute to the sustainable management of the commons. Crucial in these approaches is that they are founded on case study particularities and refrain from postulating a small set of variables that can be applied to all cases. Variables need to be chosen in response to the particular management problem under consideration. The theory offers no generalised solution that fits all cases. However, the design principles finally derived from the studies hold for a broad set of commons, although not all of them are relevant in every situation. To understand the driving forces of urban sprawl and its environmental consequences, Couch et al. (2007) proposed a set of four patterns of urban sprawl in Europe. In some of the cities investigated, several patterns apply. The patterns are associated with generalised policy recommendations.

A seminal example for an archetype analysis is the syndrome approach proposed by German Advisory Council on Global Change (WBGU 1993; Schellnhuber et al. 1997). Archetypes are identified as clusters of dynamic variables like 'urbanization', 'increasing indebtedness' or 'global warming', called symptoms. These symptoms are relatively general in nature, allowing for case-specific refinement. For example, variables like 'natural resources' can be specified as forests in one region and soil quality in another. These symptoms are related to each other through reinforcing or dampening influences. Syndromes are then introduced to disentangle this web of relations. These describe sub-dynamics of closely related trends that tend to occur jointly, such that the overall dynamics of a particular situation can be decomposed into an appropriate assemblage of syndromes. This resulted in 16 clusters, called Syndromes of Global Change, which were classified as either 'utilization', 'sink' or 'development' syndromes. The syndromes were refined by developing quantitative indicators to measure their intensity and their 'disposition' (Lüdeke et al. 1999, 2004; Kropp et al. 2001). The former refers to the actual occurrence of syndromes in a specific region; the latter to the potential for a syndrome to occur if determined factors trigger a problematic development in the future.

Using this method, two ways of defining a pattern of global change are possible: as a network of relations (which may bring about a set of scenarios), or as a set of scenarios (which may be explained by a relational network). In many modelling studies it was observed that relational networks that initially appeared as problematic can also produce positive scenarios under some conditions, making the term 'syndrome' inappropriate. It is thus currently proposed to clarify the terminology by calling the relational network an archetype. Only when the scenarios associated with the archetype are problematic are they to be denoted 'syndromes', whereas positive scenarios are called 'paradigms'. Based on this distinction, refined research questions about the conditions leading an archetype to syndromes or paradigms can be posed. It appears that the interlocking of problematic trends is a key characteristic of syndromes. If a situation gives rise to a configuration of mutually reinforcing trends, it can irreversibly 'snap' into a syndrome or paradigm (Eisenack and Petschel-Held 2002; Eisenack 2006). The analysis concentrates on clusters of mutually stabilizing trends that have the potential to bring about persistent dynamics that are judged as positive or negative according to normative specifications of the study.

Subsequently, the syndrome concept was further developed, in particular to strengthen its application for case study analysis. The Sahel syndrome and the overexploitation syndrome were investigated in more detail (Sietz et al. 2006; Kropp et al. 2006). The methods for describing syndromes were improved using qualitative differential equations (QDEs, Kuipers 1994). One interesting feature of this formal method is that it uses only the causal-loop diagram (cf. Richardson 1986) as input, such that no quantities or mathematical functions have to be defined. Theoretically speaking, a causal loop diagram subsumes a broad set of cases, where, for example, a certain dampening influence may be weaker or stronger. They thus define a functional building block that can be applied to all cases where only the so-called polarity of influences (dampening or enforcing) fits (see Petschel-Held et al. 1999; Petschel-Held and Lüdeke 2001; Eisenack 2006).

An improved approach, known under the term 'archetypes of vulnerability', was developed for assessing the vulnerability of human–environment systems to environmental and socio-economic change within the framework of the 4th Global Environment Outlook (UNEP 2007) and is being further developed in current follow-up research (Lüdeke et al. 2007; Kok et al. 2008). Archetype analysis is used to identify challenges and opportunities arising from crosscutting environmental and social processes that impact on different components of human well-being. The degree of abstraction is determined by the need for a coherent set of technical and policy options to respond to the challenges and opportunities identified. The goal is to enable responses that reduce vulnerabilities while protecting the environment. Responses are investigated with reference to a non-exclusive list of archetypes that includes global commons, contaminated sites, drylands, energy production, Small Island Developing States, technological fixes for water shortage and urbanization in coastal zones. Each description is supplied with a list of potential responses that may help to reduce vulnerabilities. It is assumed that they can be considered as potentially successful in a given situation if the basic problem description applies. The archetypes are brought about by environmental and socio-economic changes and conditions that create vulnerabilities. The problem descriptions exhibit the basic properties of archetypes: they refer to reappearing issues, for example, the atmosphere or deep-sea fisheries as global commons. They can be shared by multiple case studies, for example, rapid and poorly planned coastal urbanization in various agglomerations around the world. Finally, they are abstract descriptions to be capable of referring to generic problems. For example, technological fixes for water problems analyze the consequences of different large-scale options such as the canalization of rivers, large desalinization plants and dams. It is claimed that they can be subsumed under one archetype since successful responses can be transferred between these sub-types. The description concentrates on dams that are justified as prime examples for this archetype.

THE ACTION THEORY OF ADAPTATION

To describe adaptation to climate change in a systematic way using archetype analysis, it is first necessary to clarify how adaptation is understood in this chapter and to introduce some further concepts that facilitate more precise discussion. By restricting the analysis to adaptive behaviour by human actors, adaptation can be framed as a kind of action (see Eisenack and Stecker 2011 for a detailed exposition). Thus, established theories of action can be made use of. In analytical philosophy, action is defined to be an act, exercised by an actor with an intention (e.g., Wilson 2008). We thus propose to use the following concepts for discussing adaptation. By an impact of climate change we understand a stimulus, that is, changes of biophysical (in particular meteorological) variables triggered by climate change that affect an exposure unit, that is, actors, or social or non-human systems. An operator is an individual or collective actor that engages in adaptive behaviour. The operator employs means (e.g., material resources, legal power or social networks) to achieve certain intended ends. These ends are associated with actors, social or non-human systems, called receptors of an adaptation. Receptors may or may not be exposure units. Note that actors are *inter alia* endowed with values, beliefs, norms and means, and these attributes are important for understanding and evaluating their adaptative behaviour. The distinction between available means, employed means and necessary means for a given action can be crucial. This terminology is rooted in the 'action frame of reference' (Parsons 1937), that analyses actions by considering the actor, the ends, the situation, and the relationship between these elements. The situation is further broken down into 'conditions', which are elements the actor cannot control, and 'means', which can be controlled. The ends of actions can be defined more specifically for the purpose of this discussion, which is concerned exclusively with actions taking in response to changes in climatic conditions, by building on parts of the established terminology from the IPCC (2007b). In our analysis, the consequences of climate change are described using the concepts of state, impact, exposure unit, and response. In the context of adaptation, the responses are the actions defined previously. An impact is caused by a change in the state of biophysical conditions that affects an exposure unit which is sensitive to the change to some degree.

As a simple example, consider public information provision on risky travel behaviour with respect to, for example, heavy rain. This may be motivated by the increasing frequency and strength of extreme precipitation events. Information provision is not a concrete action, but may facilitate concrete actions to reduce potential adverse consequences of such events, in the form voluntary behavioural changes by individuals, who, for example, may select different modes of transportation in response to hearing specific weather forecasts. Here, the stimulus is the increased occurrence of extremely heavy rain, and the exposure units are users and providers of transportation. The operator is a public body that collects about changing weather conditions, and the provision of this information to the public is the means of the action. The intended end is to change behaviour of transport users, who are thus the receptors. Here, the receptors are a subset of the set of exposure units. The situation becomes more complex when we further consider the action of an administration that sets up the public body for information provision. This is a further action that should be distinguished from information provision itself. The stimulus and the exposure units are the same, but the operator is now the administration, which employs legal means and financial resources to establish the public body, which now assumes the role of a receptor. In this case the receptor is distinct from the exposure units. It is intuitively clear that the roles of operators, receptors and exposure units may be combined in a number of different ways.

It is necessary to provide some further explanation of the preceding definitions. (1) When an adaptive action directly affects the sensitivity of an exposure unit towards a stimulus, the exposure unit is also the receptor of that action. However, as the example illustrates, there can be meaningful measures that are only instrumental in the sense that they provide the means for other adaptations that would be impossible or difficult to achieve otherwise. Means and ends tend to come in chains where the effect of one action is the pre-condition for another one. Based on this, one might object that nearly every action can be classified as an adaptation, since it is not required that the receptor of the adaptation is directly affected by stimuli from climate change. I avoid this problem by using the term 'adaptation' to describe only those actions where the means-ends chain ends up at some unit that is exposed to a stimulus. (2) I use the term 'actors' to refer to both individuals and collectives of individuals. This is necessary so that the term encompasses, for example, households, companies and public bodies, all of which play an important role in societal adaptation to climate change. (3) The term 'exposure unit' has to be abstract enough to encompass the wide variety of actors, objects or systems that can be affected by climate change.

In sum, I understand adaptations as individual or collective actions, directly or indirectly intended to change the way actors or systems are affected by climate change. The proposed definitions are a formal refinement and extension of the clarifying questions "adaptation to what?", "who or what adapts?", "how does adaptation occur?" (Smit et al. 2000). The first question asks for a description of the ends of an adaptation with reference to the stimulus that affects the considered exposure unit. The second question asks for the identity of the operator, and the third requires a description on how means and ends are interlinked. A fourth question posed by Smit et al. ("what adaptations are recommended?") refers in part to the values and norms of the action frame of reference.

ARCHETYPICAL BARRIERS TO ADAPTATION

In this section I give a brief overview of the state of the art of adaptation to climate change. Based on a literature review, the archetype approach and the action theory of adaptation are applied to hypothesise about archetypical barriers to adaptation. In contrast to the expectations of economic theory already outlined, in many cases adaptive actions can only be observed to a very limited degree, while more attention is being given to mitigation activities. The literature review by Eisenack (2010) reveals that (1) many economic sectors perceive and expect increasing risks as a consequence of climate change; (2) significant mitigation activities are occurring at a local level, through governmental activities, but also in the form of local initiatives; and (3) adaptation is often neither discussed nor consciously practiced. In many cases impulses for adaptation come from government agencies and are often couched in terms of requests or targets. This is confirmed by the review of the transport sector by Eisenack et al. (2011), and is in line with the study by Ott and Richter (2008) of market opportunities for German companies arising from adaptation. For developing countries, UNFCCC (2007) concludes that although awareness of risks arising from climate change has increased, knowledge of appropriate action that can be taken in response to these risks is still rudimentary. Although there are some existing adaptation projects underway, their implementation is often difficult due to informational and capacity constraints.

These findings raise the question of what hinders adaptation to climate change at the local level. The use of the archetype approach to answer this question is motivated by some crucial characteristics of climate change and its consequences, in particular:

- Uncertainty: The type and degree of future impacts of climate change are highly uncertain due to the unpredictable effects of future global climate policies and the difficulty of downscaling global climate change to local effects, in particular with respect to extreme weather events. However, knowing about local effects is crucial for identifying appropriate local measures.
- Spatial diversity: Climate change stimuli differ greatly, even between adjacent regions. Moreover, the biophysical and socio-political conditions these stimuli will meet are very diverse as well.
- Biophysical complexity: Even within a single region there is usually a broad set of exposure units that are affected in different ways and degrees.
- Social complexity: Exposure units are interlinked in multiple ways, which means that adaptations will affect multiple receptors (not all of them exposure units) in different ways.

In short, to understand adaptation, a high diversity of local contexts needs to be considered. The archetype approach is specifically designed for such situations.

In the following paragraphs I hypothesise about potential archetypical barriers to adaptation, and describe six syndromes and one paradigm, drawing on the findings of different literature reviews (in particular Eisenack et al. 2011; Eisenack 2010; Eisenack & Stecker 2011; Hilpert et al. 2007) and a theoretical paper by Lecocq and Shalizi (2007). The current paper does not present an operationalization and validation of these archetypes, since this is ongoing and future work.

(1) Missing frames of reference: Adaptation needs are relegated to the margins of discussion about climate change. This can be characterised as a situation where no operator for adaptation exists although there may be a vague awareness of the problem. This hinders adaptation even though action is not constrained by limited means, and potential actors are not in principle opposed to adaptive action. This is caused by missing frames of reference for addressing the risks of climate change. Typically, the climate problem is framed exclusively as a mitigation challenge, and the associated risks are treated as a special class of natural hazards. In consequence, it becomes difficult to discuss, propose or even think about adaptations, since no potential operator exits. This absence may be partially explained by (a) high information costs, (b) social habits and (c) normative standards. The first reason relates to the uncertainties and spatial diversity indicated previously, which make it difficult to appreciate the need for action, even though the climate is already changing. Social habits can induce indolence if relatively new phenomena such as climate change are considered from established frames of

reference that may no longer be valid. Normative standards in the public debate on climate change reflect a strong focus on justice, as in the form of the 'polluter pays' principle. This results in a higher priority being given to mitigation, even where it is recognised that it might be a matter of responsibility to also engage in complementary adaptation activates. Public information provision and adaptation mainstreaming in sectoral policies may help to overcome the barriers to adaptation in this archetype.

(2) Moral hazard: Risky projects are undertaken in defiance of better knowledge. This archetype can explain why in many cases settlements are built or rebuilt in high-risk areas, for example, that are prone to flooding. Investors push for planning decisions in their favour, since they expect compensation from public funds for any damage arising from flooding. Municipal authorities, whose priority is local development, are often prepared to pay compensation to prevent investors from leaving the area. Analytically, the exposure units (investors) are different from the potential operators, namely local authorities. The former expect to be receptors of means in the form of compensation. Since the exposure units themselves would have the means to adapt as well, there is a mismatch of incentives. Appropriate operators and means are available, but not employed. In this case, barriers to adaptation could be overcome by establishing effective legal frameworks, for example building codes and clear rules on liability and appropriate care standards.

(3) Poverty traps: There is an inability to adapt even though there is high pressure to do so. There are potential operators, who are motivated because they are also exposure units, but capacity or budget constraints hinder appropriate adaptation—that is, the necessary means are not available. This lack of capacity to adapt increases exposure to damage caused by climate change, which further increases budget and capacity constraints, trapping actors in a vicious circle. This archetype is of crucial importance, in particular, for many developing countries that are disproportionably exposed to climate change and already have limited capacities to cope with other severe stresses. The risk may be most severe for people already living on marginal land. In this case, the means need to be made available, be it through external support (with brings its own problems) or appropriate local or regional institutional arrangements to improve access to missing resources.

(4) Mismatch of responsibilities: The power to act is not assigned. Although there are multiple exposure units and actors that could assume the role of operators, there is no actual operator of adaptation, since they all see the responsibility lying with other actors. It is unclear who has the duty or legitimacy to be the operator. There are complaints about deficient horizontal coordination between sectoral policies, such that different actors may well blame each other for the problem. Alternatively, attempts may be made to shift responsibility to higher institutional levels, since local agencies consider they do not have the means to take effective action, due to severe constraints that affect local capacity for adaptation. In a third version of this archetype, institutionalised information flows or discretionary

power between different organisations that are working well in situations which they have long experience of, do not work well in response climate change. Here, measures to restructure information flows and redefine institutional responsibilities may be helpful.

(5) Positive externalities: Less is done than it would be possible to do. Actors may choose to under-adapt when a part of the costs of under-adaptation can be shifted to others. Consequently, the necessary means are not employed even though they are available to the operator. This is due to the existence of multiple exposure units that would all be positively affected by an action taken, although only one of them (the operator) is intended to be the receptor of the adaptation. An example would be privately owned infrastructure, where the owner would incur high fixed costs if action were taken to improve its resilience to extreme weather events. But when infrastructure fails, the damage which occurs affects not only the infrastructure provider but also its users (e.g., the disruption of rail transport has costs both for the railway company and for travellers). Since it is difficult to price in fixed costs into the service offered to travellers, the infrastructure provider has no incentive to take comprehensive measures. The role of the public agencies is crucial in removing barriers to adaptation in this archetype by requiring appropriate care standards, subsidies or regulation.

(6) Clash of interests: Local conflicts arising from negative externalities of adaptations hinder decision-making. With multiple operators, the necessary means may be not employed although they are available, because proposed adaptations have multiple receptors and possibly costly side effects. For example, land use zoning policies that aim to improve coastal protection against sea level rise and storm surges require changes in land-use, such as the resettlement of people living on flood plains. These are additional costs of adaptation, that may be borne by actors who are not themselves exposure units. In this case, there is a mismatch between the willingness and power to adapt, and the legitimacy of the operator may be called into question. Depending on the institutional setting this may result in the inability to obtain majority support for adaptation measures in local negotiation processes. Such problems may be resolved by trust-building or participatory approaches, but also by creating new agencies with the power and legitimacy to act, and by making side payments to compensate negative externalities.

(7) From the bottom and the top: A fit between local agents of change and support from higher institutional levels promotes adaptation. This is a paradigm, not a syndrome like the previous archetypes described. Local operators understand the need for adaptation and can identify receptors, but insufficient means are available locally. These are provided by operators at the regional or national level. An example would be highly motivated land-use planners that do not have the budget or capacities to make or implement informed adaptation decisions, but are provided with financial and tailored information about regional climate change by the government.

Possibly new laws are passed that enable these planners to prohibit settlements in risky places.

Of course, this list is not complete and the archetypes described are not mutually exclusive. It is a starting point for further research that suggests a promising approach for disentangling and classifying the broad diversity of barriers and incentives to adaptation that are observed in practice.

CONCLUSIONS

This chapter introduced the archetype approach as a method to abstract from multiple and diverse case studies towards a comprehensive set of patterns at an intermediate level of generality. As prime example in this paper, this approach is applied to the understanding of barriers and catalysts of adaptation to climate change. Based on a definition of adaptation to climate change derived from an action theory of adaptation, a set of archetypical syndromes and paradigms (with currently hypothetical status) is presented. It is based on a qualitative survey on the state of the art of adaptation to climate change and its limitations.

The application of the archetype approach is justified by the broad variety of local and regional social-ecological systems that need to be considered to understand adaptation to climate change. In contrast, mitigation policies to reduce greenhouse gas emissions seem to be structured in a simpler way; in this case, the configuration of the society-nature interface can be described with reference to the principal emitters of a small number of relevant greenhouse gases. The impacts of climate change are, by contrast, more diverse and entangled. It seems that there are no simple instruments or policy measures that could be taken to achieve an adequate level of adaptation. There is no established common measure of a 'degree of adaptedness', and there are serious doubts about whether it will ever exist (Klein et al. 2003; Lecocq and Shalizi 2007). This could lead us to the sceptical conclusion that there can be no general adaptation policy and the associated problems have to be solved independently for every case. The archetype approach offers an approach for formulating at least weak generalisations that provide criteria for the transfer of policies from on situation to another. A transferable policy does not represent a panacea, but may facilitate the task of adapting to climate change by enabling learning from similar examples. The patterns described in this paper are intended to provide an illustration of this claim.

Of course, the diversity of social-ecological systems is not only crucial for adaptation to climate change, but for other sustainability problems as well. There is no generally applicable instrument for natural resource management under real-world conditions, but structurally similar natural resources in similar social contexts may be amenable to similar solutions. Although there are no two ecological systems with completely identical features, there may be some that share enough properties to be governed in comparable ways. For example, the social and ecological conditions and drivers of poverty may be too diverse to be tackled using only one policy instrument, unless it is not formulated in such an abstract way that it becomes meaningless. But there may be recurring patterns of poverty that can be addressed by using similar policies. I hope that the presentation of the archetype approach in this publication motivates further use of the concept in the multiple contexts where it may be applicable.

ACKNOWLEDGMENTS

I wish to thank Gerhard Petschel-Held (1964–2005), Jürgen Kropp and Matthias Lüdeke for path-breaking groundwork and general support on the issues in this paper. Diana Reckien, Esther Hoffmann and Carsten Walther contributed to literature surveys, Vera Tekken with empirical work and Micha Steinhäuser with conceptual models. All of them played their role in discussing and developing the action theory of adaptation in its current version.

REFERENCES

Cities Alliance (2007) World mayors and local governments launch climate change agreement at Bali, Indonesia (online: www.citiesalliance.org/publications/-homepage-features/dec-07/bali-agreement.html, 14.7.2008)

Couch, C, Petschel-Held, G, Leontidou, L (2007) *Urban Sprawl in Europe*. Blackwell, Oxford

Eisenack, K (2006) Model ensembles for natural resource management, PhD Thesis, Free University Berlin, DARWIN

Eisenack, K (2010) Die ökonomische Rahmung der Adaptation an den Klimawandel. In: Voss, M (ed) *Der Klimawandel—Sozialwissenschaftliche Perspektiven*. VS-Verlag, Wiesbaden

Eisenack, K, Lüdeke, M, Kropp, J (2006) Construction of archetypes as a formal method to analyze social-ecological systems, Proceedings of the Institutional Dimensions of Global Environmental Change Synthesis Conference

Eisenack, K, Petschel-Held, G (2002) Graph theoretical analysis of qualitative models in sustainability science. In: Agell, N, Ortega, JA (eds) *Working Papers of 16th Workshop on Qualitative Reasoning*. Edición Digial@tres, Sevilla, pp 53–60

Eisenack, K. und R. Stecker (2011): A framework for analyzing climate change adaptations as actions, Mitigation and Adaptation Strategies for Global Change, DOI: 10.1007/s11027-011-9323-9.

Eisenack, K., R. Stecker, D. Reckien und E. Hoffmann (2011): Adaptation to climate change in the transport sector: a review of actions and actors, Mitigation and Adaptation Strategies for Global Change, DOI: 10.1007/s11027-011-9336-4.

Encyclopædia Britannica (2006) Archetype. Retrieved November 12, 2006, from Encyclopædia Britannica Online: www.britannica.com/eb/article-9355719

Hilpert, K, Mannke, F, Schmidt-Thomé, P (2007) *Towards Climate Change Adaptation Strategies in the Baltic Sea Region—Developing Policies and Adaptation Strategies to Climate Change in the Baltic Sea Region*. Geological Survey Finland, Espoo

IPCC (2007a) *Climate Change 2007: Climate Change Impacts, Adaptation and Vulnerability, Intergovernmental Panel on Climate Change.* Cambridge University Press, Cambridge

IPCC (2007b) Summary for policymakers. In: *Climate Change 2007: Impacts, Adaptation and Vulnerability. Contribution of Working Group II to the Fourth Assessment Report of the Intergovernmental Panel on Climate Change.* Cambridge University Press, Cambridge

Jung, CG (1954) *Von den Wurzeln des Bewußtseins. Studien über den Archetypus.* Rascher, Zürich

Kelso, JAS (1997) *Dynamic Patterns.* MIT Press, Cambridge, MA

Klein, RJT, Schipper, EL, Dessai, S (2003) Integrating mitigation and adaptation into climate and development policy: Three research questions. Tyndall Centre for Climate Change Research Working Paper 40

Kok, M, Jäger, J, Karlsson, SI, Lüdeke, MKB, Mohammed-Katerere, J, Thomalla, F, Dalbelko, GD, de Soysa, I, Chenje, M,Filcak, R, Koshy, L, Martello, ML, Mathur, V, Moreno, AR, Narain, V, Sietz, D, Naser Al-Ajmi, D, Callister, K, De Oliveira, T, Fernandez, N, Gasper, D, Giada, S, Gorobets, A, Hilderink, H, Krishnan, R, Lopez, A, Nakyeyune, A, Ponce, A, Strasser S, Wonink, S (2008) Vulnerability of people and the environment: Challenges and opportunities. Background Report Chapter 7 of GEO-4, UNEP/Netherlands Environmental Assessment Agency

Kropp, J, Eisenack, K, Scheffran, J (2006) Marine overexploitation: A syndrome of global change. In: Sonak, S (ed) *Multiple Dimensions of Global Change.* TERI Press, New Delhi, pp 257–284

Kropp, J, Lüdeke, MKB, Reusswig, F (2001) Global analysis and distribution of unbalanced urbanization processes: Favela syndrome. GAIA 10: 109–121

Kuipers, BJ (1994) *Qualitative Reasoning: Modelling and Simulation with Incomplete Knowledge.* MIT Press, Cambridge, MA

Lambin, EF, Geist, HJ, Lepers, E (2003) Dynamics of land-use and land-cover change in tropical regions. Annual Reviews of Environment and Resources 28: 205–241

Lane, DC, Smart, C (1996) Reinterpreting 'generic structure': Evolution, application and limitations of a concept. System Dynamics Review 12: 87–120

Lecocq, F, Shalizi, Z (2007) Balancing expenditures on mitigation of and adaptation to climate change: An exploration of issues relevant to developing countries. Policy Research Working Paper No. 4299, World Bank

Locke, J (1690) *An Essay Concerning Human Understanding.* Book 2, Chapter 31

Lüdeke, MKB, Moldenhauer, O, Petschel-Held G (1999) Rural poverty driven soil degradation under climate change: The sensitivity of disposition towards the Sahel Syndrome with respect to climate. Environmental Modeling and Assessment 4: 295–314

Lüdeke, MKB, Petschel-Held, G, Schellnhuber, HJ (2004) Syndromes of global change: The first panoramic view. GAIA 13: 42–49

Lüdeke, MKB, Sterzel, T, Walther, C, Kok, M, Lucas, P (2007) A methodology for quantifying patterns of vulnerability. Project Report, PIK and Netherland Environmental Assessment Agency

Nordhaus, WD (1990) Slowing the greenhouse express: The economics of greenhouse warming. In: *Setting National Priorities.* The Brookings Institution, Washington, pp 185–211

Ostrom, E (2007) A diagnostic approach for going beyond panaceas. PNAS 104: 15181–15187

Ostrom, E (1990) *Governing the Commons: The Evolution of Institutions for Collective Action,* Cambridge University Press, Cambridge

Ott, HE, Richter, C (2008) Anpassung an den Klimawandel—Risiken und Chancen für deutsche Unternehmen. Wuppertal Papers 171, Wuppertal Institut

Parsons, T (1937) *The Structure of Social Action.* McGraw Hill, New York
Petschel-Held, G, Lüdeke, MKB (2001) Integration of case studies on global change by means of qualitative differential equations. Integrated Assessment 2: 123–138
Petschel-Held, G, Block, A, Cassel-Gintz, M, Kropp, J, Lüdeke, MKB, Moldenhauer, O, Reusswig, F, Schellnhuber, HJ (1999) Syndromes of global change: A qualitative modelling approach to assist global environmental management. Environmental modeling and assessment 4: 315–326
Ragin, CC (1987) *The Comparative Method.* University of California Press, Berkeley
Richardson, GP (1986) Problems with causal-loop diagrams. System Dynamics Review 2: 158–170
Schachhuber, A (2004) *Social Movements, Environmental Governance, and Rural Territorial Development: An International Perspective*, Rimisp (Centro Latinoamericano para el Desarrollo Rural)
Schellnhuber, HJ, Block, A, Cassel-Gintz, M, Kropp, J, Lammel, G, Lass, W, Linienkamp, R, Loose, C, Lüdeke, MKB, Moldenhauer, O, Petschel-Held, G, Plöchl, M, Reusswig F (1997) Syndromes of global change. GAIA 6: 19–34
Shrader-Frechette, KS, McCoy, ED (1993) *Method in Ecology: Strategies for Conservation.* Cambridge University Press, Cambridge
Sietz, D, Untied, B, Walkenhorst, O, Lüdeke, MKB, Mertins, G, Petschel-Held, G, Schellnhuber, HJ (2006) Smallholder agriculture in northeast Brazil: Assessing heterogeneous human–environmental dynamics. Regional Environmental Change 6: 132–146
Smit, B, Burton, I, Klein, RJT, Wandel, J (2000) An anatomy of adaptation to climate change and variability. Climatic Change 45: 223–251
UNEP (2007) Vulnerability of people and the environment: Challenges and opportunities. Chapter 7 of *Global Environmental Outlook 4.* United Nations Environment Programme, Nairobi
UNFCCC (2007) *Climate Change: Impacts, Vulnerabilities and Adaptation in Developing Countries.* Climate Change Secretariat, Bonn
Urpelainen, J (2009) Explaining Schwarzenegger: Local frontrunners in climate policy. Global Environmental Politics 9: 82–105
Warren, A (2002) Land degradation is contextual. Land Degradation and Development 13: 449–459
WBGU (1993) *World in Transition: Basic Structure of Global People-Environment Interactions. German Advisory Council on Global Change, Annual Report.* Economica Verlag, Bonn
Weber, M (1922) *Gesammelte Aufsätze zur Wissenschaftslehre.* Tübingen
Wilson, G (2008) Action. In Zalta, EN (ed) *The Stanford Encyclopedia of Philosophy* (online: http://plato.stanford.edu/archives/fall2008/entries/action/, 14.7.2008)
Windelband, W (1919) Geschichte und Naturwissenschaft (Straßburger Rektoratsrede 1894). In: Windelband, W (ed) *Präludien, Aufsätze und Reden zur Philosophie und ihrer Geschichte II.* Tübingen, pp 136–160
Wittgenstein, L (1953) *The Philosophical Investigations*, §§ 66
Wolstenholme, EF (2003) Towards the definition and use of a core set of archetypal structures in system dynamics. Systems Dynamics Review 19: 7–26

7 Population Dynamics and Adaptive Capacity of Supply Systems

Diana Hummel

INTRODUCTION

In his notes about population and environmental services the Canadian scholar Vaclav Smil states that there seems to be no escape from reductionism in modern science.

> Even those subjects that should retain some semblance of holistic approach—if not being the very paragons of it—end up being chopped up into pieces that may be convenient for standard intellectual transactions but are distortions of not just vastly more complex, but also highly dynamic, realities. And so we have population and food, population and water, population and climate, population and whatever . . . as if any of these subjects could be treated independently of many other factors, that is, as if there were no complex links and feedbacks among them. (Smil 2002: 187)

These considerations aptly describe the difficulties of the relationship between the 'abstract' scientific world and the 'concrete' world of everyday phenomena (see Becker, this volume). They also point to the link between normative and analytical dimensions challenging social-ecological systems (SES) analysis in general, and thus also studies of the population–nature–society nexus.

The point of departure for the theoretical approach discussed in this chapter is the normative assumption that a changing number of people in a given society implies regulatory problems for structures that supply basic goods and services such as energy, water and nutrition, resulting in social-ecological problems. The central argument is that it is not population dynamics *per se* that generates these problems, but rather the inadequate capacity of supply systems to adapt to and cope with demographic changes. Against the background of the societal and scientific problems resulting from the lack of adaptive capacity, an integrated model will be developed for analysing the interactions among population, nature and society.

POPULATION DYNAMICS AND SOCIAL-ECOLOGICAL TRANSFORMATIONS

The Millennium Ecosystem Assessment (MEA 2005) includes the development of human population among the primary, interconnected driving forces of global ecosystem change. Physical and biological drivers such as changes in climate, plant nutrient use, land conversion, diseases and invasive species are considered to be direct drivers that unequivocally influence ecosystem processes. Indirect, anthropogenic drivers operate more diffusely by altering one or more direct drivers. In addition to demographic changes, anthropogenic drivers include economic, socio-political, cultural and religious, scientific, and technological transformations. Changes in ecosystem services are caused by the interactions among these multiple drivers over time. For example, population and income growth interact with technological advances to induce climatic changes. The drivers interact across spatial, temporal and organisational scales, and the combined effects of multiple drivers are amplified by reciprocal action and feedbacks (cf. Nelson et al. 2006). As of yet, however, these interactions among anthropogenic and biophysical drivers are only partially understood.

Viewed through the lens of SES analyses, population dynamics can be conceived as a central attribute of social-ecological transformations. Over the past few decades, the world has witnessed tremendous demographic changes. At the global level, the demographic transformations are characterised by trends such as an ever-growing population, although at a slowing pace; declining fertility rates; increasing migration and urbanization; as well as rising life expectancy and further demographic aging expected in the coming decades (Population Reference Bureau 2008; United Nations 2007; Lutz et al. 2008; Hummel, Hertler, Niemann, et al. 2008; Cohen 2003). While these transformations appear to be general and universal demographic patterns at the global level, population dynamics have produced unprecedented demographic diversity across regions and countries. In addition to the overall trends, it is the disparity and incongruity of demographic changes that represent essential characteristics of global population dynamics. These changes become apparent between as well as within individual regions and countries, such as is the case with respect to changes in fertility rates and their pace, or alterations of the population's age structure. The result is that traditional demographic groupings are transforming. For example, many regions in developing countries like Brazil, India and Thailand now have the low fertility rates that until recently were only found in high-income countries, while in many African countries the decline in fertility has ground to a halt. In contrast, average life expectancy has decreased due to the precarious economic and social conditions in many Eastern European transition countries[1].

Studies on demographic drivers of (global) environmental change generally concentrate on numerical variables of the primary determinants population change—fertility, mortality and migration—and changes in the population size and distribution. This implies something of a reductionist perspective of the reciprocal impacts of ecosystem functions and population dynamics. From a social-ecological perspective it must be emphasised that demographic changes and population dynamics consist not only of quantitative processes such as changing fertility or mortality rates and numbers of inhabitants in a given area, but also of phenomena such as altering household and family structures, lifestyles, income, education level, social status and consumption patterns (see Wildenberg and Singh, this volume). Demographic changes are thus embedded in social-ecological transformations. One example is urbanization; as a worldwide social phenomenon, it is connected with changing needs and habits, higher incomes and a rise in living standards, combined with changing individual values, preferences and demands. As a result of urbanization and globalization, Western lifestyles are becoming more attractive around the globe, which in turn alters consumption behaviour and the related effects on ecosystems.

Different and sometimes contradictory problem descriptions regarding the diverse population dynamics can be found in both the scientific and the public discourse. On the one hand, population growth in developing countries is still regarded as a major cause of ecological degradation including rising emissions, desertification, deforestation and the overuse of natural resources. On the other hand, arguments about the negative effects of population growth on the environment and welfare in developing countries, characteristic of the sustainability debate until the end of the twentieth century, are being increasingly displaced by discussions about demographic changes in more developed countries. Instead of growing populations, it is *de*creasing population size and aging societies, and the respective impacts particularly for economy (e.g., labour market, social welfare systems, family structures, etc.), that dominate the discussion about demographic change in most Western societies.

POPULATION–ENVIRONMENT ANALYSIS[2]

In recent decades, research on population and environment has advanced institutionally and intellectually, spanning different scientific disciplines, in particular demography, geography, economics, and anthropology. This emerging interdisciplinary field of studies is frequently designated as 'population–environment (P-E) analysis'. Because of its interdisciplinary nature, there exists a variety of theoretical perspectives and methodological approaches within the field. P-E analysis aims at drawing a comprehensive picture of the relations between population and environment by addressing the 'cycle of interactions'. At issue is not only the question "What are the effects of changes in the human population on the natural environment (P-E)?", but also the converse query, "What are the effects of changes in the natural environment on the human population (E-P)?" (Lutz, Scherbov, Prskawetz, et al. 2002: 225).

The Malthusian view of a linear, causal relationship between population growth and degradation of ecosystems has been largely discredited in the scientific debate (Smil 2001; Cohen 1995). Current P-E literature (de Sherbinin et al. 2007; Hunter 2000; Curran and de Sherbinin 2004; Lutz, Prskawetz, et al. 2002; Panayotou 2000; Arizpe et al. 1994) reveals that direct causal explanations oversimplify complex realities and are thus of little instructive value. The result has been a shift in the P-E literature from a concern for the depletion of non-renewable resources such as fossil fuels and minerals to an interest in the dynamics of household decision-making *vis-à-vis* land use, fertility, migration, and livelihood strategies. More emphasis is also given to 'intervening conditions' affecting the population–environment relationship, and P-E researchers have thus begun to incorporate social, cultural, and institutional factors that 'mediate' between 'P' and 'E' in their models. This 'mediating variable approach' stresses that the relationships between population dynamics and the environment are highly dependent on a range of contextual factors such as macro-economic policies, globalization and production of exportable resources, and institutions governing resource access, as well as local or region-specific dynamics. It should be noted that all these factors belong to the category 'society'. Meanwhile there is growing recognition in the research field that

> populations are composed of people who collectively form societies, and people and societies cannot easily be reduced to food and material demands that result in some aggregate impact on the environment. (de Sherbinin et al. 2008: 363)

This perception is accompanied by an increasing impact of system-theoretical approaches in the population, development, environment (PDE) research field. System-theoretical perspectives view environment and population as interacting (complex) systems. One of the most advanced models in P-E analysis is the International Institute for Applied Systems Analysis' (IIASA) 'PDE model'. This model has been adopted in various empirical case studies and thus continues to evolve (Lutz, Sanderson, Wils, et al. 2002). Because of the issue of food insecurity in African countries, the agriculture (A) factor has been added, resulting in a PEDA model. As Figure 7.1 illustrates, the PEDA model links population parameters (sex, age, migration, etc.) to other, in the narrow sense non-demographic, but rather socio-economic variables such as education and gender-specific labour force, to issues such as land degradation and food production and distribution (Lutz and Sherbov 2000).

The combined field of study has advanced considerably in recent years. Specific dimensions of P-E analysis, such as interactions and feedbacks, spatial and temporal scale, complexity, vulnerability, and uncertainty, among others, are also the subject of SES analysis. Accordingly, Lutz, Prskawetz,

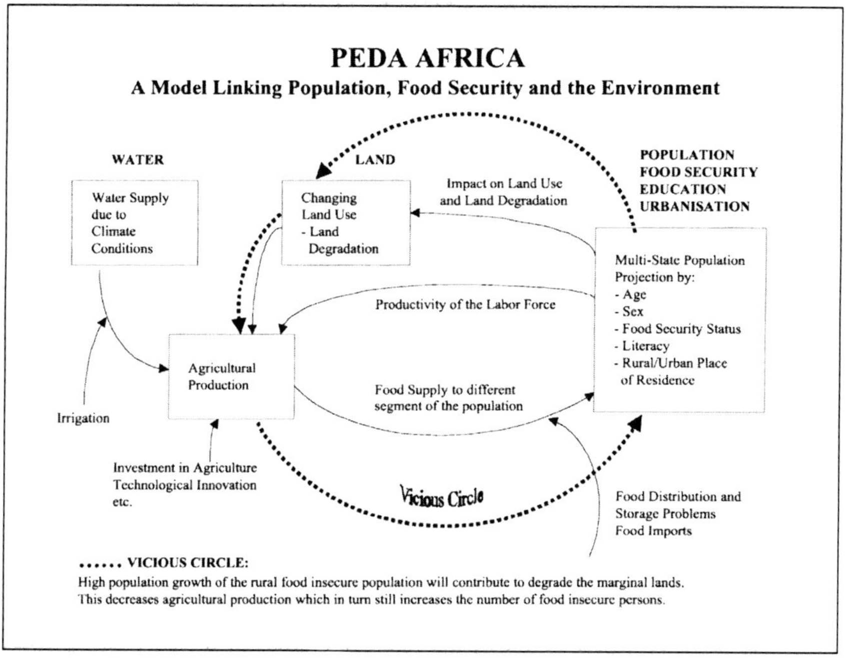

Figure 7.1 Basic structure of the PEDA model (Lutz and Sherbov 2000). With kind permission from Springer Science+Business Media: Essays in Honor of Gustav Felchtinger, "Optimization, Dynamics and Economic Analysis," 2000, p. 313, Engelbert J. Dockner and Richard F. Hartl (eds.), Fig. 1).

et al. (2002: 5) describe P-E research a chair with four legs: 1) P (population dynamics), 2) E (environmental dynamics), 3) influences of P on E, and 4) influences of E on P. However, to borrow from the metaphor, since the focus in P-E analysis is on the first three legs, the chair is slightly wobbly. Indeed, it is commonly understood in P-E analysis that population dynamics impact social, cultural, political, economic, and ecological development, and, *vice versa*, that demographic processes are influenced by social, cultural, economic, and ecological conditions. There are, thus, causal relations at work. But going even further, a reciprocal relationship must be assumed in the sense of circular causality; in turn, the effects of the dynamics and the bundle of factors influence the initial causes, making a clear identification of cause and effect no longer possible. Small deviations in the initial conditions can alter the course of events and produce large changes that can result in new patterns of relations between nature and society. Thus, models that can map this complexity, including the consideration of positive and negative feedback loops, are needed[3]. The PDE model is a step in this direction but exhibits a number of shortcomings regarding the 'environment' category. For example, in P-E analysis 'environment' is subdivided into 'natural' and 'human made';

into different ecological compounds like soil, water, and so on; and into different environmental problem fields such as natural resource depletion, air or coastal pollution, and so on. Clearly, in order to describe issues such as contamination, degradation of soils and resources, and so on, the term 'environmental problems' is indispensable in its everyday use. Yet, from an analytical point of view, problems arise by virtue of the *relational character* of the notion 'environment'. 'Environment' always refers to the environment of something. People do not relate to their environment as isolated individuals, but as social individuals and hence as civil subjects within specific cultural, social, and economic conditions, for example, by nurturing children and elders, practicing agriculture, producing goods and services, or doing science (Hummel 2000: 308ff.; Becker and Jahn 2006: 142f.). Thus, humans as well as institutions maintain relations to their specific environment. The following section illustrates an alternative to P-E research.

SUPPLY SYSTEMS AS SOCIAL-ECOLOGICAL SYSTEMS

The social-ecological approach attempts to avoid both the naturalist and the socio-centric perspective and seeks to pursue a critical and integrated view. It relates demography to the interactions between 'nature' and 'society', focusing on interactions between demographic processes and social-ecological problems (Hummel, Hertler, Niemann, et al. 2008; Lux et al. 2006). One basic assumption is that demographic changes are at any time and at any place connected to social-ecological problems that cannot be assigned to one or the other of two exclusive categories 'nature' or 'society'. The cognitive interest is thus not restricted to the effects of demographic changes on either the environment or society. Population changes depend on biophysical conditions, as well as on social, cultural and economic settings, and in turn they influence these factors. In this sense, demographic changes—be it rising or falling population numbers and fertility rates, changes in age structure, migration, urbanization processes, or household structures—indicate transformations *of societal relations to nature*, that is, the relational network formed by individuals, societies and nature in interaction (Becker and Jahn 2006).

Demographic changes and the conditions under which they take place are connected with the nexus of societal relations to nature such as work and production, food and nutrition, mobility, or housing. The analysis of population dynamics in the context of social-ecological transformations needs therefore to be connected with the issue of the adaptation of societies to demographic changes, the development capacities of societies, and forms of social-ecological regulation. A novel approach is to systematically link demographic changes to the issue of provision. Every society must address the task of providing its population with food, energy, water, housing, education and health services in such a way that simultaneously meets basic human needs, ensures a decent quality of life and preserves the natural life-supporting functions of ecosystem.

Adopting the SES perspective on the interactions of supply systems and population dynamics, however, renders the issue even more complex. Instead of seeking for the relevant demographic factors and their environmental outcomes, or studying the conditions for 'sustainable population development', the *non-sustainability* of provisioning is stressed by focusing on the social-ecological transformations and regulations of supply systems and their adaptive capacity for coping with demographic changes.

By modelling the dynamic relation patterns between the natural and the social sphere, provisioning structures developed by societies to satisfy the basic needs of the population can be conceptualised as SES. They cover biophysical and material-energetic dimensions (e.g., technical artefacts such as wells or bridges) as well as cultural-symbolic aspects of life (e.g., gender and social roles, needs, values, attitudes, cognitive orders). The 'social' and the 'natural' are linked in certain, context-specific ways and develop specific problem dynamics in which social, economic, technical, political, and ecological problems closely interact. In a nutshell, supply systems are characterised by a coupling of natural and social elements, which together, in their interactions and corresponding sets of problems, engender specific societal relations to nature. They are regulated by societies, and, at the same time, they are dependent on natural conditions. This kind of SES can be structured and simplified as follows (Figure 7.2):

A major particularity of this model is that users are understood to be an integral part of supply systems. The term refers to actors and actor constellations, includes both providers (producers) and receivers (consumers) of supply system services, and can distinguish between direct and indirect use

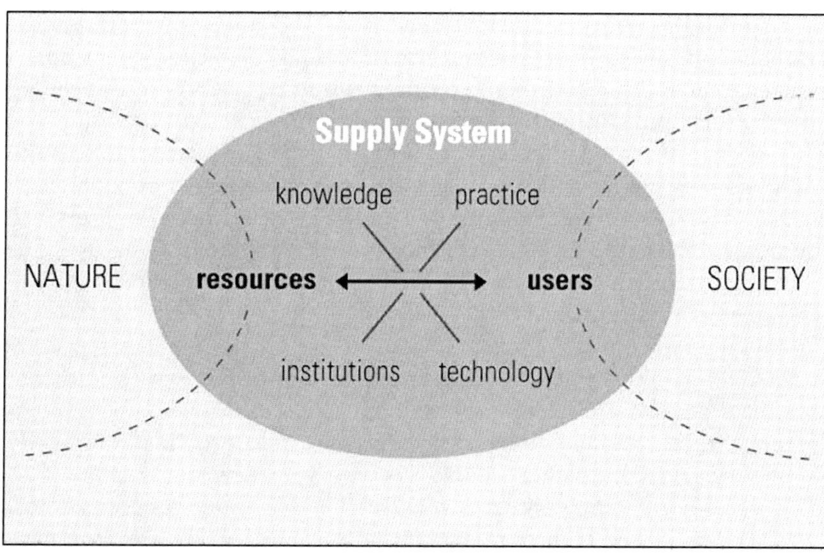

Figure 7.2 Supply systems as social-ecological systems (Hummel et al. 2008).

of resources. The users of supply systems are not identical with a population; instead, each group of users must be analyzed for specific supply systems. For instance, user groups of water supply systems usually encompass various components such as individuals, households, public water utilities, industry and agriculture. Depending on specific supply systems, differentiations must be made between and within different user groups. These differentiations can also highlight competition among different user groups and purposes of resource use (e.g., cultivating crops for human consumption, as animal feed, or as bio-fuels), as well as conflicts between societal use of resources and preservation of ecosystem functions.

The process of resource utilisation is determined by knowledge, practices, institutions and technology. These terms describe ways and means of resource availability and allocation and determine the vulnerability, adaptivity, scope and options of regulating provision. *Knowledge* comprises both scientific and expert knowledge as well as common sense knowledge. *Practices* represent systematised types of behaviour that encompass forms of bodily and mental activities, practical activities and their representations, as well as their interactions. The term includes both social, discursive practices and material ones, all of which are carried out by various societal, specifically situated actors. *Institutions* represent socially established rules of action, informal constraints, and formal rules that structure political, economic, and social interactions and provide a framework for the action. *Technology* comprises all material structures designed, built and controlled by humans for achieving specific purposes, including physical infrastructures, logistics and other technical elements used by producers or consumers of services. These intervening dimensions are related to one another. For example, cultivation forms are embedded in specific socio-cultural, economic, political and institutional contexts including practices such as gender-specific labour division, knowledge about appropriate cultivars, and availability and application of technical equipments, financial resources, legislation, subsidies, and so on. Depending on the regional and cultural context and kind and purpose of provision, the specific relevance and relation between knowledge, practices, technology and institutions must be identified. In sum, the internal dynamic of supply systems gives rise to specific configurations of the relations between users and resources. At the same time, it influences societal relations to nature. The relationships between society and nature transform as the relations between users and resources change.

Today parts of modern supply structures are spatially and temporally flexible and thus highly mobile. Due to globalisation, production and consumption processes are becoming interconnected worldwide. Technological change and economic globalisation have led to a growing complexity of supply systems. Individual supply systems are increasingly coupled to one another. Food supply systems require effective water supply systems, as well as established transport and traffic systems, which are nowadays increasingly dependent on energy supply, as well as information and communication networks. Supply systems are constituted and interact at various spatial scales—globally, regionally and locally. An increasing number of supply system elements are being connected by technological means (crops, farmers, transport enterprises, power plants, internet, etc.). In this way, individual supply systems spread out spatially, increasingly encompassing the entire globe, and the number of elements and nodes integrated in the system and interacting with one another grows. This results in increasing possibilities of mutual influence, dependency and feedback loops. Evermore connected global production and distribution chains combined with rising specialization increases the susceptibility to disturbances.

It goes without saying that demography and provision are mutually dependent. Changes in numbers, distribution and social composition of the population in a given area influence the spatiotemporal dynamics of supply systems. Demographic processes influence supply systems, and in turn the organisation of a supply system has impacts on demographic development. These interactions give rise to specific challenges for the adaptivity and regulation capacity of supply systems.

CASE STUDY EXAMPLE: URBANISATION AND FOOD SUPPLY IN GHANA

How do demographic changes and transformations of supply systems structurally relate to and interact with one another? What are specific regulatory and adaptation problems? The following case study[4] focuses on the issue of urbanisation as one major feature of population dynamics and provides empirical findings on which to base answers to these questions.

In the international discussion of the relationship between demographic development and food supply, the phenomenon of urbanization has played an increasingly important role in recent years. The beginning of the 21st century has been accompanied by a dramatic social-spatial reorganisation of the world's population. Now, for the first time in the history of humankind, more people live in urban areas than in rural ones, and developing countries in particular have displayed a historically unique degree of urbanisation. Specific *urban societal relations to nature* take shape with this process, and these can undergo crisis-prone forms of development, such as poverty, environmental degradation and food insecurity, resulting in high morbidity and mortality. These phenomena are particularly true for African urban areas, but they are not restricted to the 'black continent' (Janowicz 2008a, 2008b; Hummel and Janowicz 2006).

Viewing the problem of food supply from the perspective of supply systems permits a change of perspective in which the food crisis is seen not in terms of a 'disturbed equilibrium' between population and resources, but rather as being caused by a lack of adaptivity on the part of urban supply systems resulting from dynamics of social-ecological crisis. This lack of adaptive capacity becomes evident, for instance, when unsuccessful regulation of urban agriculture results in precarious land and property relations in land, leading to changing environmental conditions, for example, to less and less available fertile land which in turn results from competitions between different ways of land use and overuse.

African cities face enormous challenges including desolate housing conditions, lack of sanitary facilities, massive ecological degradation, high unemployment and precarious water and food supply situations. Ghana provides a good example of critical developments in urban food supply systems resulting from urbanisation processes. The country is one of the most dynamic regions of West Africa, and today the degree of urbanization stands at around 40%. Accra, the capital, with a population of approximately 1.8 million, has more than doubled in size over the last 20 years. During the same time its Metropolitan area has increased by 318%. The share of people working in agriculture is high, and food supply is hardly secure. Twenty-five percent of children are undernourished, 24% of households face food insecurity, and 40% of households are considered vulnerable (Janowicz 2008b: 142f). Urban agriculture (inner-city farming, house gardening, and peri-urban farming) is a significant factor in assuring the security of food supplies in Accra; however, as a result of urbanization this practice has increasingly come under pressure. Land necessary for farming is becoming scarcer as urban growth from the city centre outwards destroys the surrounding fields that could be otherwise used for farming. Huge areas of this land are lost every year, either directly due to housing construction or indirectly due to the extraction of construction materials (e.g., sand, stones). Thus, different uses—economic (agriculture, industry and trade), political (administration and military) and social-cultural (residents)—compete for the same functional spaces.

In this regard, institutional factors are crucial. The situation in Ghana is marked by a lack of clarity concerning legal regulation of property relations in urban areas, which aggravates the competition for land use. Institutional rules, policies and procedures relating to rights of ownership and use play an intrinsic role by setting up specific resource regimes that consist of a mixture of traditional, religious, everyday practical and administrative arrangements. It is especially the existing legal pluralism with respect to property rights in land that gives rise to different forms of legitimate ownership or use claims that exist simultaneously. The result is a heterogeneous order of fragmented urban areas marked by legal insecurity and unpredictability in the application law. One strategy to cope with this uncertain legal order is changing actions and techniques tied to a given form of farming. This can include shifting cultivation or intercropping, reducing fallow periods, or planting crops with a short growth cycle. In turn, these processes result in over-fertilization and nutrient depletion, leading in turn to harvests with less quantity and of poor quality. The vicious cycle is intensified both by the high loss factor in foodstuffs transported from farms far from urban areas due to insufficient infrastructure and by the lack of almost any form of urban planning that could minimise expansion into ecologically sensitive areas. This, in turn, has economic consequences such as higher prices for produce. The destruction of urban farmlands leads to a lengthening of transport routes and the costs of this are passed on directly to consumers. At the beginning of the period of rapid urbanisation and strong country-to-city migration in Ghana, legal indeterminacy strengthened urban agriculture and enabled a successful adaptive strategy. Today, however, this legal pluralism threatens the adaptability on the part of the food supply system. In sum, the case study revealed that it is not demographic processes and urbanization processes by themselves that lead to food insecurity, but their combination with crisis-ridden transformations of societal relations to nature, resulting in precarious forms of food supply.

ADAPTIVITY AND REGULATION CAPACITY OF SUPPLY SYSTEMS

The case studies carried out in the *demons* project indicated that the transformation of supply systems for water and food, and the crisis-prone developments they undergo, are not solely or primarily caused by population dynamics; instead, such dynamics mesh with specific context-determined societal and natural conditions. Nevertheless, significant common patterns among the various case study regions could be identified with respect to the interactions of demography and transformation of supply systems, as well as common challenges for the their adaptivity and regulation capacity (Hummel, Hertler, Janowicz, et al. 2008: 221ff.).

- Increasing or decreasing population size clearly influences the structures of demand for provisioning goods and services, and thus the size of a population plays an important role in the transformations of supply systems.
- However, size does not constitute the sole explanatory variable to water and food consumption; other variables such as income, education, household structures and social or kinship networks also play a role. For example, shifts in household structures are often associated with changes in population distribution and settlement patterns.
- Migration turned out to be the most relevant driving force for the transformations of supply systems. Improvements in a supply situation can function as an attractor, or pull-factor, promoting the spatial population movements. For instance, the expectation that better and more secure provisioning services will be found can be seen as an important motive for migration to urban centres. Migration processes are tied to changing population distribution and settlement patterns, and because of their short-term character (compared to other demographic processes such as changes of fertility or life expectancy), they

> present a particular challenge for the adaptation and regulation of supply systems. This is primarily because they are associated with strong fluctuations in demand and use of resources and are, at the same time, extremely difficult to predict.

We thus see the mutual reinforcement of population dynamics and transformations of supply systems. What is more, demographic changes represent a special challenge for predictions and estimates concerning future developments because of their non-linearity. The longer time span of a prediction, and the smaller the area is valid for, the larger the degree of its uncertainty, since ostensibly incidental changes in basic parameters can lead to significant effects for a supply systems. For example, unexpected developments such as technological change or the modification of usage patterns as well as climate factors cause insecurity that make precise water demand forecasts almost impossible.

This problem is aggravated by the superposition of long- and short-term dynamics, both within demographic dynamics themselves, as well as within the interplay of demographic changes and transformations of supply systems. First, short-term population dynamics (e.g., migration) and long term demographic changes (e.g., life expectancy) overlap and result in temporal-spatial fluctuations. Uncertainty is also connected to the interaction of short- and long-term dynamics in demographic trends with changes in demand and in infrastructure development. As with population, supply systems have their own dynamics, and there is a lack of spatiotemporal congruence between supply structures (and their transformation) and population dynamics. Against this background, the adaptive capacity and sustainable regulation of supply systems faces the challenge of dealing with the 'typical' complexity of SES, that is, the interconnectedness of social, ecological and technical elements and processes; the multiplicity of scales; non-linear interactions; self-organisation; and uncertainty.

Sustainability relating to supply systems is difficult to determine in general, as it always depends on specific regional and historical contexts, as well as on specific cultural conditions. This paper thus seeks not to formulate a concrete set of instructions for political, economic or societal practice, but rather to illustrate certain properties of the regulation capability of supply systems. Borrowing from the 'guiding principles for a sustainable resource management' proposed by Kluge et al. (2006, 2007), these interrelated properties can be described as adaptivity, transformativity, integration and functionality.

Adaptivity refers to the ability to adapt to changing targets within a process; in the supply systems context it can be concretised as the latter's capacity to adjust to changed initial conditions. Thus, adaptivity is far from passive reaction. An important presupposition is the productive use of existent feedback loops linking consequences and effects to decisions made and measures taken. It particularly refers to loosening the grip of path dependencies and related inertial tendencies.

Transformativity implies the ability of a system to form new structures and processes and to develop further. In terms of supply systems this means the ability to provide goods and services reliably, as well as the ability to respond to changes in demand in a timely manner. It includes quantitative aspects such as a growing or shrinking demand for particular goods and services, as well as qualitative aspects like changes in the quality requirements for supply services, and spatial aspects such as the emergence of new geographical centres of increased or diminished demand.

Integration refers to the fact that resource management decisions pertain to different kinds of expertise, practices, and institutions and affect different groups of actors, which have to be integrated in processes of decision-making. At the core of this participatory approach is the need to expand the knowledge base available to all interested parties and to develop a common understanding of the problem at hand. Alternative options for action have to be assessed and conflicting interests must be highlighted. This can help to built trust and raise the level of acceptance for decisions and implementation of measures on the part of the various actors. A prerequisite is to include those groups whose demands, values, and claims will be affected by the consequences of any decisions made.

Functionality refers to a supply system's economic prerequisites to its managerial requirements, that is, to ensure the interruption-free fulfilment of its goals and quality standards. Functionality is also a basic presupposition of sustainability, both in terms of the functionality of ecosystem services and of societal provisioning structures. Ecological, economic and social aspects must be combined during the implementation process of specific measures and includes enhanced management and evaluation paradigms that are geared to uncertainty, complexity of structural conditions governing supply, and in particular feedback effects. This can be implemented, for instance, in scenarios or in monitoring processes.

The most important element of all kinds of policies is an awareness of the different effects actions might have both on ecological conditions, as well as on social structures such as on the interests of resource users, and on economic conditions. An understanding of these effects in terms of feedback loops is necessary so that such feedback loops can be taken into account when putting together further, possibly altered, regulatory measures. Social-ecological regulations become in this way iterative, adaptive learning processes.

CONCLUSION

Societal, ecological and demographic developments are strongly interdependent, yet remain undetermined and open. Boundary condition population dynamics significantly influences societal relations to nature. Yet demographic changes alone do not determine societal development and its natural basis because societies can actively shape the impacts of demographic changes through sustainable provisioning of goods and services. In contrast to the usual view of population dynamics as drivers for crisis-prone transformations, they can be then regarded as critical forces for more sustainable development paths.

The conceptual model of supply systems allows an analysis of the interactions of population, society and nature in a novel way, a way in which different demographic developments are systematically related to the issue of provisioning and patterns of resource use. As has become clear, interactions among demographic changes and transformations are characterised by process dynamics shaped by feedback loops and network structures, making population dynamics and supply systems a co-evolutionary set, both temporally and spatially. It seems appropriate then to model this co-evolutionary set as a complex dynamic system. Further development of the model of supply systems towards stronger formalization remains, however, a task for the future.

ACKNOWLEDGMENTS

This chapter could not have been written without intense discussion and cooperation. It draws on research carried out within the interdisciplinary project demons—demographic trends, needs and supply systems, financed by the German Ministry of Education and Research (BMBF) within the program "Social-Ecological Research" SOEF from 2002 through 2007. I would like to express my sincere thanks to my colleagues and members of the demons team Christine Hertler, Alexandra Lux, Cedric Janowicz and Steffen Niemann.

NOTES

1. Given these divergent trends and patterns of population dynamics, demographic transition theory's claim of universality can be put into question.
2. The following section refers in part to Hummel et al. (2009).
3. Feedback can take two forms: as a strengthening of an effect (positive feedback) or as a counterbalancing of an effect (negative feedback).
4. The case studies carried out in the *demons* project focused on countries and regions with strong population dynamics (e.g., South-East Asia, Middle East, Namibia, Ghana, eastern Germany) and thus covered multiple scales (national, sub-national, regional). Each case study concentrated on specific demographic aspects (e.g. migratory movements, population growth, or decreases in population size) and either on water or food supply (Hummel 2008).

REFERENCES

Arizpe, L, Stone, P, Major, D (1994) *Population and the Environment: Rethinking the Debate*. Westview Press, Boulder

Becker, E, Jahn, T (2006) *Soziale Ökologie. Grundzüge einer Wissenschaft von den gesellschaftlichen Naturverhältnissen.* Campus, Frankfurt, New York

Cohen, JE (1995) *How Many People Can the Earth Support?* W. W. Norton & Company, New York, London

Cohen, JE (2003) Human populations: The next half century. Science 302: 1172–1175

Curran, SR, de Sherbinin, A (2004) Completing the picture: The challenges of bringing 'consumption' into the population–environment equation. Population and Environment 26: 107–131

de Sherbinin, A, Carr, D, Cassels, S, Jiang, L (2007) Population and environment. Annual Review of Environment and Resources 32: 345–373

de Sherbinin, A, VanWey, LK, McSweeney, K, Aggarwal, J, Barbieri, A, Henry, S, Hunter, LM, Twine, W, Walker, R (2008) Rural household demographics, livelihood and the environment. Global Environmental Change 18: 38–53

Hummel, D, Lux; A, de Sherbinin, A, Adamo, SB (2009) Theoretical and methodological issues in the analysis of population dynamics and supply systems. Background Paper for the Population-Environment Research Network (PERN) Cyberseminar on "Theoretical and Methodological Issues in the Analysis of Population Dynamics and the Environment", 2–13 February 2009. www.populationenvironmentresearch.org

Hummel, D (2000) *Der Bevölkerungsdiskurs. Demographisches Wissen und politische Macht.* Leske & Budrich, Opladen

Hummel, D (2008) *Population Dynamics and Supply Systems: A Transdisciplinary Approach.* Campus, Frankfurt, New York

Hummel, D, Hertler, C, Janowicz, C, Lux, A, Niemann, S (2008) Part C: Synthesis. In: Hummel, D (ed) *Population Dynamics and Supply Systems: A Transdisciplinary Approach.* Campus, Frankfurt, New York, pp 211–259

Hummel, D, Hertler, C, Niemann, S, Lux, A, Janowicz, C (2008) Part A: The analytical framework. In: Hummel, D (ed): *Population Dynamics and Supply Systems: A Transdisciplinary Approach.* Campus, Frankfurt, New York, pp 9–69

Hummel, D, Janowicz, C (2006) Demografisierte Konflikte. In: *informationszentrum 3.* Welt iz3w: 33–36

Hunter, LM (2000) *The Environmental Implications of Population Dynamics.* RAND, Santa Monica

Janowicz, C (2008a) *Zur Sozialen Ökologie urbaner Räume. Afrikanische Städte im Spannungsfeld von demographischer Entwicklung und Nahrungsversorgung.* transcript, Bielefeld

Janowicz, C (2008b) The world goes urban: Food supply systems and urbanization processes in Africa. In: Hummel, D (ed) *Population Dynamics and Supply Systems: A Transdisciplinary Approach.* Campus, Frankfurt, New York, pp 129–160

Kluge, T, Liehr, S, Schramm, E (2007) Strukturveränderungen und neue Verfahren der Ressourcenregulation. ISOE-Diskussionspapiere 27. Frankfurt am Main, ISOE

Kluge, T, Libbe, J, Scheele, U, Schramm, E, Trapp, JH (2006) Der netWORKS Ansatz zur integrierten Strategiebildung. In: Kluge, T, Libbe, J (eds) *Transformationen netzgebundener Infrastruktur. Strategien für Kommunen am Beispiel Wasser.* Difu, Berlin, pp 33–56

Lutz, W, Prskawetz, A, Sanderson, WC (2002) Introduction. In: *Population and Environment: Methods of Analysis.* Special Supplement to the Population and Development Review 28: 1–21

Lutz, W, Sanderson, W, Scherbov, S (2008) The coming acceleration of global population ageing. Nature 451: 716–719

Lutz, W, Sanderson, WC, Wils, A (2002) Conclusion: Comprehensive P-E studies. In: *Population and Environment: Methods of Analysis.* Special Supplement to the Population and Development Review 28: 225–250

Lutz, W, Scherbov, S (2000) Quantifying vicious circle dynamics. The PEDA model for population, environment, development and agriculture in African countries. In: Dockner, EJ, Hart, RF, Luptacik, M, Sorger, G (eds): *Optimization, Dynamics, and Economic Analysis. Essays in Honor of Gustav Feichtinger.* Physica, Heidelberg

Lutz, W, Scherbov, S, Prskawetz, A, Dworak, M, Feichtinger, G (2002) Population, natural resources, and food security: Lessons from comparing full and reduced-form models. In: *Population and Environment: Methods of Analysis.* Special Supplement to the Population and Development Review 28: 199–224

Lux, A, Janowicz, C, Hummel, D (2006) Versorgungssysteme. In: Becker, E, Jahn, T (eds) (2006) *Soziale Ökologie. Grundzüge einer Wissenschaft von den gesellschaftlichen Naturverhältnissen.* Campus, Frankfurt, New York, pp 423–433

MEA—Millennium Ecosystem Assessment (2005) *Ecosystems and Human Well-Being. Synthesis.* Island Press, Washington, DC

Nelson, GC, Bennett, E, Berhe, AA, Cassman, K, DeFries, R, Dietz, T, Dobermann, A, Dobson, A, Janetos, A, Levy, M, Marco, D, Nakicenovic, N, O'Neill, B, Norgaard, R, Petschel-Held, G, Ojima, D, Pingali, P, Watson, R, Zurek, M (2006) Anthropogenic drivers of ecosystem change: An overview. Ecology and Society 11: 29 (online)

Panayotou, T (2000) Population and environment. In: Tietenbert, T, Folmer, H (eds) *The International Yearbook of Environmental and Resource Economics 2000/2001.* Edward Elgar Publishing Ltd., Cheltenham, Northampton, pp 148–196

Population Reference Bureau (2008) 2008 World Population Data Sheet. Washington, DC

Smil, V (2001) *Feeding the World: A Challenge for the Twenty-First Century.* MIT Press, Cambridge, MA

Smil, V (2002) Population and environmental services. Population and Development Review. Supplement to Vol. 28: 187–198

United Nations (2007) World Population Prospects. The 2006 Revision. Highlights ESA/P/WP.202. New York

Smil, V (2001) *Feeding the World: A Challenge for the Twenty-First Century.* MIT Press, Cambridge, MA

Smil, V (2002) Population and environmental services. Population and Development Review. Supplement to Vol. 28: 187–198

United Nations (2007) World Population Prospects. The 2006 Revision. Highlights ESA/P/WP.202. New York

Part IV

Multi-Agent Modelling and Simulation

8 Transdisciplinary Multi-Agent Modelling for Social-Ecological Systems Analysis

Achievements and Potentials

Marion Glaser

INTRODUCTION

Social-ecological systems (SES) analysis focuses on human–nature dynamics which generate—or have the potential to resolve—sustainability problems. It proposes an integrated view of complex human–nature relations, a paradigm shift.

SES definitions vary for different types of analysis (see Becker, Tretter, Halliday and Glaser et al. p. 199 this volume). For the purposes of this chapter, an SES is conceptualised as consisting of interlinked natural and social hierarchies as initially proposed by Berkes and Folke (1998). While alternative, more holistic systems views have other advantages (Halliday and Glaser 2011), the conceptualization of social and ecological subsystems allows for a collaborative interdisciplinary investigation of the linkages between those components. This complements the analysis of their internal workings (Gunderson and Holling 2002; Glaser 2006) and provides the basis for connecting the different rich bodies of relevant disciplinary knowledge which are growing in the natural and social sciences in relative isolation.

Thus, for instance, where a spatially defined natural system (e.g., a local reef, a mangrove coastal region or river drainage basin) with an identified human–nature problem (e.g., destructive fishing or marine pollution) is the starting point for social-ecological analysis, the question is: How do interacting social and ecological drivers—from various levels in the earth system—generate the identified sustainability problem at the defined focal level? The SES researcher (team) now needs methods to undertake such analysis.

Thus far, methodological options are scarce. New concepts for the investigation of social-ecological change such as emergence, resilience, adaptive and transformative capacity have established themselves and gained in theoretical depth in recent years (Holling and Gunderson 2002; Holling et al. 2002; Olsson et al. 2004; Cumming 2011). The combination of disciplinary methods to address social-ecological problems in large multidisciplinary team building and analysis exercises has been advanced through pioneering work such as that of Rosemary Ommer and her research team (Ommer 2007). Arguably, however, the development of integrated methods in which disciplines collaboratively develop methods of reconstructing the emergence of complex system outcomes is still largely outstanding (Glaeser et al. 2009). In this context, multi-agent modelling[1] has recently been described as a "preferred tool of the emerging field of complexity science" (Shafer 2007). The remainder of this chapter explores this.

Some SES definitions incorporate the agency concept. Lansing and Kremer (1993:110) speak of networks of interacting agents (processes, elements) exhibiting dynamic aggregate behaviour that emerges from the individual activities of the agents. Anderies et al. (2006) define SES as composed of agents ranging from microbes to plants to humans, each with a different degree of information-processing capacity. This author's SES definition (Glaser et al. 2010) adopts a focus on stakeholders from any level of the global social-ecological complex as system agents which may be combined with the analysis of natural dynamics, including the behaviour of non-human living agents.

With a particular focus on coastal and marine social-ecological systems (CM-SES), this chapter analyses how agent-based modelling measures up to what is required from a sustainability-oriented SES analysis.

CM-SES are located in the catchment to coast plus open sea region. On the ecosystem side, they are characterised by specific bio-geochemical features (e.g., upwelling, sandy, rocky, estuarine or lagoon, pelagic or sea bottom) and ecosystem types (e.g., coral, seagrass, mangrove and other wetlands). Their specific institutional and social characteristics are particularly diverse use patterns, mobile user populations, often undefined boundaries and territories as well as particularly weak and less structured governance systems[2]. These features generate special problems for analysis. Mobile user populations of CM-SES—such as migrant fisherfolk and boats/industrial ships—and the occurrence of many non-territorial species in marine environments complicate boundary definitions both in the horizontal and in the vertical direction. Perhaps also because human–nature interactions are more easily observed in terrestrial than in coastal and marine systems, multi-agent models have so far mainly focused on terrestrial systems. The need to feed system knowledge into the design of governance structures is therefore particularly pronounced for coastal and marine social ecological systems.

WHAT DO WE REQUIRE FROM A METHOD TO ANALYSE SOCIAL-ECOLOGICAL SYSTEMS DYNAMICS?

Before discussing multi-agent modelling in the following section, we need to be clear what to expect from it as an adequate method for social-ecological analysis. I suggest that any method for sustainability-oriented social-ecological analysis will need to perform well in at least four arenas:

1. *It should deliver new knowledge.* The collection and interpretation of data should be rigorous, transparent and offer well-founded, plausible arguments about the significance of data (Spencer et al. 2003). Sets of information which vary in exactness, resolution as well as temporal and spatial scales will need to be combined.
2. *It should identify the drivers of complex social-ecological systems dynamics.* Major questions here are as follows: How do desirable and undesirable system features emerge? Which rules produce sustainable resource use under which circumstances? Which options exist for different decision-makers? What kinds of futures are likely? The emergence of social-ecological features and system capacities such as resilience, vulnerability, adaptivity, and transformability (Capra 2007; Grimm et al. 2005) needs to be understood and reproduced at different temporal, spatial and institutional scales (Cash et al. 2006). Governance and management can then be better designed to support sustainable human–nature relations.
3. *It should integrate knowledge from different disciplines.* Effective communications between the separate conceptual universes of the social and natural sciences and the humanities are needed to develop holistic and problem-focussed approaches to sustainability issues in the human–nature realm.
4. *It should support the active participation of ecosystem users*[3] (Arnstein 1969; Govan 2009). Ecosystem users' individual and collective learning capacities need to be strengthened to generate knowledge for supporting sustainable social-ecological dynamics. To grasp the complexities surrounding human behaviour and associated social ecological feedback loops, knowledge sharing beyond the academic disciplines is needed. Ecosystem users have central stakes in "their" systems and often possess outstanding knowledge about SES features within their range of experience (Johannes 1981). The integration of this knowledge into analysis is crucial (Berkes and Folke 2002; Nicolson et al. 2002) to enable ecosystem users to influence research priorities, to contribute to analysis and to co-design solutions.

Points 3 and 4 combine into a call for transdisciplinary practice in SES analysis, where transdisciplinarity describes all forms of interdisciplinary cooperation to generate knowledge which involve not only scientists but also practitioners from beyond the realm of science (e.g. resource users) in research (Defila and DiGiulio 1999).

INTERDISCIPLINARY MULTI-AGENT MODELLING

Multi-agent modelling starts with the individual system actor. It examines individual agency, its causes and effects and traces how system changes feed back into individual agency. This approach is part of a "top-down to bottom-up" paradigm shift in diverse disciplines such as ecology (Uchmanski and Grimm 1996), rural sociology and planning (Chambers 1992), demography (Billari et al. 2006), complex dynamical systems theory (Emmeche et al. 1997) and modelling (Barreteau 2003).

Agent Types

For the purposes of SES modelling, there are three agent types (Janssen 2002: 3):

- Humans who differ in mental maps, goals, locations, abilities and endowments
- Non-humans such as animals and plants
- Passive agents such as chemical particles and other non-living entities

In natural science models, agents can be living or non-living "ants in a colony, neurons in a neural network, particles in physics" (Holland 1998: 123). In social models, an agent may be a human individual, household or an organisation (village community, firm, fishing vessel), a living individual or a composite social entity.

Purposive agency assumes that agents have specific objectives and resources (Anderies 2002; Gebetsroither 2006). The objectives of human agents can be investigated. Animals and even plants can be treated as if they had objectives. In contrast to this, non-living entities, however, do not have objectives in the sense of "an own agenda".

Agent behaviour is the driver of multi-agent models (Pyka and Grebel 2006). As modellers, Pereira and Reis (2004) define an agent as "a self-contained software program, specialised in achieving a set of goals, by autonomously performing tasks." Living agents display goal-oriented behaviour according to simple rules in reaction to their immediate environment. Humans are more complicated than other agent types. They rationalise, act and react in accordance with dynamic values, perceptions and interpretations.

In interdisciplinary multi-agent modelling, an agent is a discrete natural or social entity situated in space and time. Agents display individual combinations of attributes; their individual life history may affect behaviour; they may interact (compete, exploit, cooperate) at specified system levels; and learn, adapt or reproduce in accordance with their own state and that of their environment[4].

Agent Hierarchies and Scales within Nested Systems

SES and other multi-scale systems are hierarchies, characterised by asymmetrical relationships between system levels. The position of an agent within these hierarchies affects its information level and scope for action. The spatial manifestations of linked nested biophysical and social hierarchies do not usually coincide, however. This generates the "problem of fit" and associated problems (Folke et al. 1998; Cash et al. 2006).

At higher system levels, a social entity (e.g., a household, organisation or population) that is defined by a set of rules and contains a number of smaller agents may be defined as an agent. In multi-level agent-based modelling, aggregate agents can thus consist of and emerge from lower-level agents.

What Is Multi-Agent Modelling and What Are the Main Procedural Steps?

A computer-based multi-agent model simulates the actions and interactions of individual system agents in order to assess their combined effects on system configurations and functions. Self-organisational capacities are assumed for the system so that phenomena such as cooperation in resource management, resource degradation or sustainable management may emerge from the rule-based interactions of system agents. Iterative multi-agent modelling relies on repeated model runs to explore under which conditions which system features and patterns emerge. This is an important asset for scenario development.

Step 1: Identify research problem. Model objectives need to be specified. Which and whose questions will be addressed? Possible purposes of a model are the reconstruction of observed real life patterns, the exploration of possible futures or the didactic presentation of system dynamics, system understanding or societal learning. Variable overload may occur or target objectives conflict unless it is clear what and who the model is for. The exclusion of potential model users from this initial step can cause non-acceptance of the product (Nicolson et al. 2002; Shafer 2007).

Step 2: Design model. Agent-based SES models can embed social agents into bio-geochemical environments, focus on the interactions of human and non-human agents, or a combination of both. Modellers need to decide on system boundaries, identify important agents and determine their objectives and the rules for agent (inter)actions. Decisions on "agent architecture" are needed such as: Are agents communicating? Are they reflexive? Do they learn? Which features of their environment do they react to? How do they react? Over-complexity, with its multiple error potentials, and undue simplicity, which may omit major system drivers are potential pitfalls. The minimum number of variables to answer the model purpose should be included.

Step 3: Construct model. There is a choice between programming and using one of the available packages or toolkits. Software packages tend to be easier to use but programming gives more scope to address

Continued

the unexpected. An overview of this rapidly developing field can be found at http://www.swarm.org/index.php/Main_Page[5].

Step 4: Test model functioning. Ensuring that the modelling program functions without crashing or generating internal errors is the next step, a technical task for the modeller.

Step 5: Ground-truth the model. Validation 'ground-truths' the goodness of fit of model structure and outputs with the model's empirical referents (Manson 2002). It can be undertaken with secondary data that was not used as model input, with experiments or via participatory simulation. Participatory role games can simulate real-life situations to test model assumptions about the rationales, norms and behaviours of agents in the model.

Step 6: Interpret model results. Experts (including local stakeholders) debate the relevance, realism and implications of model outputs. Problem redefinitions might lead to model redesign and a new modelling cycle.

Box 8.1 summarises the major procedures in multi-agent modelling. The presented sequence of six steps may include circular loops, for instance between model construction and ground-truthing.

To link knowledge systems and to capacitate ecosystem users for sustainable system management, transdisciplinarity is essential. In multi-agent modelling this means that interdisciplinary models need to be developed, ground-truthed and interpreted in a participatory, inclusive manner. In the following review, we illustrate some achievements of and challenges for multi-agent modelling in SES analysis.

MODELLING EMERGENCE: SEPARATE BEGINNINGS

Effective SES analysis needs to understand and reproduce the emergence of social-ecological system features and capacities[6]. Both the natural and the social sciences have investigated emergence through agent-based modelling.

In the social sciences, agent-based models illustrate how heterogeneous individual behaviours arising from a variety of simple rules at the micro-level generate societal outcomes. An early example of this bottom-up approach to modelling, Epstein and Axtell's (1996) "Sugarscape", is an entirely artificial 'virtual world', in which the authors trace the emergence of migration, pollution, sexual reproduction, combat and culture to the rule-based (inter) actions of individual social agents. Other multi-agent models of social systems have investigated how heterogeneous individual actions affect a social system's structure and behaviour (e.g., Axelrod 1984 on the emergence of cooperation), and how changes in societal structure influence the actions of individual agents (Sawyer 2005). The cognitive agents in agent-based models have heterogeneous knowledge, beliefs and goals. This renders rigid assumptions about human behaviour—such as in the infamous *homo oeconomicus*[7] concept—unnecessary. Gilbert and Troitzsch (2005) introduce a range of agent-based social models which are based on the behaviour of households. These simulate, for instance, the impact of tax reforms, gender segregation and opinion formation. Such simulation of social dynamics in "virtual experiments" with artificial societies (Sawyer 2005) allows heterogeneity in values and behaviours among system agents and has provided new insights on social emergence[8].

In the natural sciences, Grimm and Railsback (2006) show how individual-based modelling can simulate the emergence of patterns in ecosystems. By testing whether models of alternative forms of adaptive rule-based behaviour by "non-human living agents" (i.e., animals and plants) reproduce patterns that are also observable in reality, the authors construct a virtual laboratory for complex ecosystems analysis. The results of such virtual experiments are more "structurally realistic" when the model-based reproduction of real-life patterns succeeds at several system levels.

SES analysis requires the integrated analysis of the kind of social and ecological dynamics reproduced in the previously cited disciplinary models. This creates a need to engage with both the social and the natural drivers of the behaviour of human and non-human system agents. To the authors' knowledge, most individual-based modellers in ecology (Grimm and Railsback 2006 and references under http://www.abm-ecology.org.references) and agent-based modellers from the social sciences (Gilbert and Troitzsch 2005; Ferrand 2007 and publications on companion modelling) have so far stayed within their own disciplinary boundaries.

EXPLAINING SYSTEMS DYNAMICS: A FIRST INTERDISCIPLINARY PHASE

Interdisciplinary multi-agent models of SES are novel tools and they mostly deal with land use and planning (Janssen 2002; Berger and Schreinemachers 2006[9]). Agent-based models of CM-SES are particularly rare. Although Janssen (2002) observes that rule-based models of resource appropriation by individual agents can explain a range of "puzzling historical observations", few interdisciplinary models are validated with reference to specific real-life systems. Even fewer include system stakeholders in model development, validation or analysis (Janssen and Ostrom 2006). In the following, seven interdisciplinary models of social-ecological dynamics are briefly introduced.

1. In one of the rare attempts at multi-agent modelling of a CM-SES, Rudd et al. (2002) outline the spatial components of a model for conch management for a 68 km^2 area, the East Harbour Lobster and Conch

Reserve in the South Caicos Islands. The agents in this model are the conch fishers and a fishery manager. Informed by Landsat images, the model simulates conch feeding, growth, habitat selection and mortality in grid-type habitat cells. Model assumptions about ecology and habitat are supported by site-specific data. Fishers are conceptualised as simple profit maximisers. The fishery manager in the model has a number of policy options for conch management (e.g., to vary total allowable catch or minimum size limits).

2. Huigen et al. (2006) focus on the identification of differing human rationales for land use decisions. Their model's objective is the understanding of interacting demographic and land use changes in the San Mariano watershed in the Philippines. The approach relies on socio-economic data and anthropologically derived 'real-life stories'. These are transformed into an agent-based model according to a set procedure. Model outputs are then validated against local narratives and statistical data. The model reveals the influence of ethnic identity on land use behaviour.
3. In a module-based multi-agent model, Shafer (2007) simulates how an integrated "coral reef-fish-fishermen SES" responds to different management approaches. This 'synthetic' model does not relate to any specific geographic region. It simulates the characteristic dynamics of a coral-based and fisheries-reliant SES as a specific archetype (see Eisenack, this volume) and generates emergent phenomena such as fish schooling, resource depletion, information diffusion and cooperation.
4. The model on lobster fishery on the coast of Maine by Wilson et al. (2007) and Yan et al. (2008) investigates how lobster fishers decide where to place their traps and whether to destroy the traps of other fishers if the chance arises. The model has a biophysical and an agent-based, social part. It assumes dynamic environmental conditions such as seasonality, spatial heterogeneity of lobster distribution, sea bottom type, water depth and temperature. Three spatial scales with their respective system drivers are simulated. Agents (fisherfolk) in the model "learn" according to a transparent, computable pattern, their decisions evolve with the success of previous fishing efforts. Model outputs show the emergence of territoriality and social structure. Validation with a four-year data set which includes one million trap hauls shows that the model reproduces actual cooperation and conflict patterns in its geographic area of reference. Under conditions of agent learning, the model generates the emergence of territorially based fishing groups, as they had indeed developed in the geographic reference area.
5. Castella et al. (2005) model land use as the outcome of social-ecological interactions arising from the decisions of individual human agents. With research data, role-playing games and GIS, they develop a transdisciplinary model of a multi-scale SES with focus on a mountain

region in Vietnam. Their participatory multi-agent model is designed as a communication and learning tool. Their iterative approach to model validation with ecosystem users who have little formal education[10] derives complex social-ecological patterns from simple rules which are evaluated and revised with local system users. This is done in a participatory process which covers model construction, groundtruthing and interpretation.

6. Focusing on the objectives of collective learning and didactic effectiveness, Cleland (2007)[11] analyses the processes associated with and outcomes of two participatory agent-based modelling workshops with local system stakeholders in Mexico and the Philippines. The workshops apparently created a space for reflection and interaction: In a "fun virtual world", coral ecosystem users altered the rules of the multi-agent model on sustainable reef management to reflect their realities, perceived options and understandings. Workshop participants contrasted their own learning processes in this exercise positively with the more common use of local people as "informants" from whom data is extracted for analysis elsewhere. Participation also altered model assumptions: Workshop participants had criticised that the initial model contained only economic criteria for agent behaviour. In order to align local perceptions with model assumption, risk minimizing and satisficing strategies and "other understandings of aspirations, place and identity" were incorporated into later versions of the model.
7. Gurung et al. (2006) report on nine years of design, implementation and impact assessment of a multi-agent model of water use in a watershed in Bhutan. At the outset, the study region was ridden with severe water distribution conflicts. Role plays—in individualist and collective modes of decision-making under varying environmental and social conditions—are at the centre of this iterative modelling approach. The authors stress that long-term involvement by researchers and facilitators in the research region is a precondition for the realisation of the achieved empowerment and conflict resolution potentials of participatory multi-agent modelling.

DISCUSSION AND OUTLOOK

Is Multi-Agent Modelling an Adequate Method for SES Analysis?

Multi-agent modelling has multiple potentials for social-ecological analysis. It uses transparent procedures to reconstruct the major causal elements associated with the complex and inherently unpredictable emergence of social-ecological patterns. It can increase our understanding of system dynamics and emergence, combine data types and provide options to validate model outputs against 'real-life' data. It can also function as a learning and communication tool, provide inputs for governance or resolve conflicts and empower system users. In some of the previously cited examples, multi-agent models provide an interface between natural and social science, or between political decision-makers and the otherwise often marginalised priorities and knowledge of ecosystem users. That this can lead to the empowerment of system users, that is, to the strengthening of individual and organisational capacities for cooperative learning and planning has also been shown (Gurung et al. 2006). Participatory multi-agent models have improved stakeholders' understanding of the social-ecological dynamics they are part of and supported individual and collective decision-making (Bousquet et al. 2002; Castella et al. 2005; Cleland 2007).

Critics argue that, with many relevant variables and even more feedbacks between them, multi-agent models are too technically complex, too demanding on the individual researcher and too data intensive. While the likelihood of surprise and uncertainty may also have increased, advancing computer capacities and ever better data banks render such technical complexity manageable now (Janssen et al. 2008; Wilson et al. 2007).

Anderies (2002) cogently remarks that "stakeholders in resource systems seldom want to `experiment' with their livelihoods". The virtual laboratory character of multi-agent modelling eliminates the ethical and practical difficulties of the experimental approach in the world of living, sentient beings. Moreover, computational laboratories which use the multi-agent approach can—and are starting to explore possible futures in ways which are not feasible in the slower—and potentially too destructive—"learning by doing" processes in real-life situations.

Table 8.1 assesses the seven interdisciplinary SES models discussed previously according to the four criteria for effective, sustainability-oriented SES analysis proposed under the second section of this chapter.

While this assessment lays no claim to ultimate validity, it does demonstrate the diverse objectives and achievements of interdisciplinary multi-agent models in SES analysis to date.

1. The model by Rudd et al. (2002) links major resource management policy options to ecological dynamics but it lacks important detail on the social side such as the differentiation of fisherfolk decisions resource endowments or personal values and rationales.
2. The model by Huigen et al. (2006) combines demographically guided land use decisions into an agent-based behavioural model. This connects qualitative anthropological with quantitative socio-economic data, an important feat in interdisciplinary practice. With their temporally explicit spatial modelling of possible behavioural complexes of interacting agents, Huigen et al. also highlight ethnicity, a highly sensitive policy issue and driver of land use decisions in their geographic area of reference. However, validation of this model proved to be only

Table 8.1 Objectives and Performance Rating of Interdisciplinary Social-Ecological Models

Authors	Model Objectives	New knowledge	SES drivers identified?	Interdisciplinary integration?	Local participation
1. Rudd et al. 2002	Supports conch fishery management	*	*	*	–
2. Huigen et al. 2006	Understand interaction of demographic and land use changes	***	**	*	**
3. Shafer, 2007	Understand generic system type dynamics in reef fishery	***	**	***	–
4. Wilson et al. 2007; Yan et al, 2008	Explain emergence of fisheries governance	***	***	***	–
5. Castella et al. 2005	Provide communication & learning tool in land use management	**	**	**	***
6. Cleland, 2007	Collective learning in reef management	**	*	*	***
7. Gurung et al. 2006	Institution-building in watershed management	*	***	*	***

– none, * rudimentary, ** acceptable/good, *** excellent

moderately successful perhaps because variations in nature such as soil type and quality and associated dynamics were not included.

3. By combining data from different reefs, fisheries and societies around the globe into an integrated model, Shafer (2007) is methodologically innovative. Her approach to interdisciplinary multi-agent modelling clearly improves our understanding of human–coral-fish interactions understood as a specific generic type of human–nature interactions. However, perhaps because this model has no geographical area of reference, its empirical 'ground-truthing', the participation and capacitation of local system users, the incorporation of local knowledge and the generation of synergies between local and disciplinary knowledge are particularly difficult and mostly missing.
4. The model developed by Wilson et al. (2007) and Yan et al. (2008) successfully reproduces the co-evolution of environmental conditions,

local institutions and learning-based behavioural dynamics among fishermen. The authors interpret the territorial fishing patterns generated by their model as a "precursor of (effective) fisheries governance" for the study region. This model successfully simulates—but it does not stimulate—the learning of system agents. It traces the emergence of resource use rules to the interaction of social and ecological variables. Ground-truthing is successful: Model outputs correspond to real-life social-ecological patterns. These are two important achievements. However, since neither model construction nor ground-truthing include any form of system user participation, important insights/components of user knowledge may have been missed.

5. The multi-agent model which Castella et al. (2005) produced in cooperation with local system users generates patterns very similar to real-life land use patterns in the model's geographical reference area. Since it succeeds in reconstructing real-life social-ecological dynamics, this transdisciplinary model is of high relevance to local and regional governance and management. Methods to up-scale such participatory local and regional models are still needed. Additionally, ecological change needs to be better included as a dynamic and co-evolving model element in the companion modelling approach of which Castella et al.'s work is an example.
6. The incorporation of local perceptions and knowledge in the agent-based model described by Cleland (2007) is reported to have improved model quality and furthered a sense of ownership of the generated knowledge among local stakeholders. However, ground-truthing of this model was not possible since ecological scales were manipulated for didactic purposes in a target conflict between the didactic and the system analysis/knowledge generation objectives of the model(ling process). Cleland emphasises that participatory modelling efforts need to be embedded in comprehensive and locally led involvement with system users. This is a complex and long-term task. As did early PRA practitioners (e.g., Chambers 1992), Cleland also observes that the quality and policy effectiveness of the knowledge generated in participatory (modelling) exercises suffer when, as often happens, powerful decision-makers do not take part.
7. Finally, Gurung et al. (2007) document how a long-term, iterative and participatory agent-based modelling process supported knowledge generation and exchange and facilitated institution building between stakeholders with strongly divergent interests. The authors find that collective model building and validation improved communication between researchers and locals and document the resulting improvements of local capacities for sustainability-oriented decision-making. While an eminently successful tool for conflict resolution and collective learning in a complex social-ecological context, this model is weaker in terms of generating new social-ecological knowledge.

Agent-based models can integrate the different social realities perceived by heterogeneous social actors with natural science data—which may also be presented in an agent-based mode—to (re)construct and analyse integrated human–nature dynamics. The few examples to date show that transdisciplinary multi-agent models can explain counterintuitive observations and support the development of management approaches with system users.

Our initial demands for an adequate method for SES analysis are thus clearly met. The multi-agent approach can deliver new knowledge on complex SES dynamics and it can support interdisciplinarity with the active participation of ecosystem users. Key concepts in SES analysis such as co-evolution, emergence and self-organisation are being explored through multi-agent modelling with encouraging results.

However, as Table 8.1 clearly shows, the multiple potentials of multi-agent modelling have not been, and perhaps cannot be, simultaneously realised by the same model. Target conflicts between the different possible purposes of multi-agent modelling have surfaced: The model of Wilson et al. (2007) and Yan (2008) shows that the use of model-immanent learning algorithms, which requires non-interference with the research universe, is not easily combined with the capacitation of ecosystem users. The didactic purposes of the participatory model described by Cleland (2007), on the other hand, undermined that model's capacity for system analysis and knowledge generation.

An early and clear establishment of model objectives is thus essential to avoid overload and target conflict problems in multi-agent modelling. The multiple objectives of social-ecological analysis can be met by multi-agent modelling but multiple models and research approaches are likely to be needed.

Further Potentials

Given the preceding analysis of achievements to date, there are at least four promising arenas for the further development of multi-agent modelling in SES analysis:

Constructing Human Actor Types

The agent concept opens new avenues for the study of human–nature relations. The impact of agents' internal beliefs and norms on their behaviour towards nature can now be simulated in virtual experiments. Comparative field-based experiments in different parts of the world may be used to identify the impact of culture on human perceptions and uses of ecosystems[12]. The realization of the potentials of participatory agent-based modelling in this has only just begun. New results might be expected for questions such as, How do social groups and networks emerge and reproduce over time? Under what conditions do individuals put group interests before personal interest? While social theory might thus be advanced, the identification of points of leverage supporting desirable social-ecological system change is also possible.

In an important move beyond simple conceptions of 'economic man' whose behaviour is oriented solely towards short-term economic gain, Jager et al. (2001) and Jager and Janssen (2002) assume two psychological types (*homo oeconomicus* and *homo psychologicus*) whose different rationalities generate different behaviours in otherwise equal contexts (*ceteris paribus*). Their aggregation of a variety of possible psychological types as one *homo psychologicus* needs ground-truthing in different socio-economic and cultural contexts. Field experiences in poor rural environments (Lipton 1978; Glaser 2003) suggest minimisation of livlihood risks as one possible alternative rationality structure. Finding out whether, where and under which conditions different rationalities prevail can be investigated with participatory agent-based models.

Simulating the Co-Evolution of Living Human and Non-Human Agents

Social-ecological emergence cannot always be reproduced in models with relatively inflexible individual agents, which follow simple rules (Holland 1998). An understanding of complex cognitive and culturally specific processes is needed to simulate human–nature dynamics (Jager et al. 2000; De Vries, 2006). With pattern-orientation in ecological modelling (Grimm et al. 2005) and in interdisciplinary work (Wilson et al. 2007; Yan et al. 2008), efforts to reconstruct past social-ecological patterns and to simulate possible future ones in the 'virtual' worlds of spatially specific transdisciplinary multi-agent modelling have gained in effectiveness. While already successfully integrating knowledge about the behaviour of non-human living agents (i.e., trees, fish, crabs, (examples under http://www.abm-ecology.org/references and in Reuter et al. 2005; Breckling et al. 2005) multi-agent modelling also provides the scope to include deeply human characteristics such as subjectivity, reflexivity, normativity and self-conceptualisation into SES analysis. Most multi-agent models of social-ecological dynamics focus on either the social or the ecological side but rarely on both in a balanced way[13]. The modelling of interactive human and non-human living agents is still rare, and a promising option for the future.

Ecosystem User Participation in Modelling

Transdisciplinary social-ecological research which aims for participatory multi-agent modelling is rarer still. In conventional modelling, "the modeller enters into the process with (his or her own) hypotheses or idiosyncrasies" (Ferrand 2007: 209) and translates these into model assumptions. In participatory modelling, on the other hand, system users may challenge the cultural leanings of modellers. During iterative participatory model construction, the rules and norms of behaviour that agents follow in the model are scrutinised by some of those people who are part of the social-ecological complex which is the subject of the model. Participatory practice in multi-agent modelling to date has revolved around spatially based board and role games. SES users 'role play' diverse positions of human agency towards the ecosystems in question. Rules, decision-making criteria and system knowledge are iteratively reviewed as the model is designed or validated. The experiences, worldviews and knowledge of those who live with SES under study thus become part of the model. Role games directly activate cognitive processes in the players. This may reduce the need for complex cognitive models (Ferrand 2007). It needs to be investigated under which circumstances human behaviour in games differs from behaviour in real life situations.

There are numerous challenges to the effective participation of ecosystem users in multi-agent modelling. Communication in international, multi-ethnic or transdisciplinary teams requires intercultural competence. Local ecosystem users need an understanding of the modelling concept and of 'their' SES model to be able to engage with it. Communication standards and guidelines on how to describe multi-agent models clearly and in sufficient detail to enable synergies, interfaces and exchange are now being initiated by ecologists (Grimm et al. 2006) and social scientists (Janssen et al. 2008), but apparently without much communication between those groups. To enable the effective participatory modelling of SES dynamics, work already done in the PRA[14] and communications fields will need to be integrated into modelling teams and into the development process of integrated transdisciplinary modelling standards.

Developing Multi-Level Approaches

So far, interdisciplinary multi-agent models cluster at the local and regional level do little to support our understanding of larger scale systems or to address global sustainability. Methods to up-scale agent-based models or to integrate them with other models at higher system scales are needed. Biermann (2007: 328) argues that "earth system governance . . . implies some form of self-regulation of societal actors . . . in the solving of societal problems". To better understand the cross-scale effects of such dynamics, the behaviour of individual and aggregate agents at higher spatial and institutional levels needs to be clarified. The potentials of interdisciplinary multi-agent modelling are only just beginning to be realised at the local/regional level (e.g., Wildenberg and Singh, this volume). In future, network analysis may be used to conceptualise higher system levels as "interactive nodes" (Flitner and Görg 2008). The interfacing of multi-agent with other types of models is also just being conceptualised (Kroeze et al. 2008).

Multi-scale analysis also needs improved participatory practice. Hare et al. (2003) investigate how SES agents are selected for participation in modelling. They reason that to ensure model efficacy, a match between the "scale of action and influence" of the selected system stakeholders and the purposes of the model is needed. The effective crossing of spatial and institutional scales would render multi-agent modelling vastly more effective in SES analysis.

CONCLUSION

The development of multi-agent models clearly addresses the current methodological gaps in SES analysis. Much remains to be done, however. A new phase for the development of transdisciplinary social-ecological models which builds on the achievements of mono- and interdisciplinary beginnings and on the first findings of interdisciplinary and participatory modelling practice is possible now.

ACKNOWLEDGMENTS

Thanks to Broder Breckling, Gesche Krause and Hauke Reuter for their comments on previous versions of this chapter and to Karl-Heinz Simon and Andrew Halliday for their challenging and detailed reviews of this chapter.

NOTES

1. Multi-agent modelling is also referred to as agent-based modelling and, in the natural sciences, as individual-based modelling. This article uses the terms multi-agent and agent-based modelling interchangeably
2. The weak governance feature of CM-SES was identified in the conclusions of the International Workshop on the Governance of Coastal and Marine Systems, Resilience 2008 Conference Stockholm, April 18, 2008.
3. *Active* participation means the assumption of explicit rights and responsibilities in the process in question
4. This definition has benefitted from a discussion with and an unpublished paper by Flora Vadas in 2008.
5. Thanks to Hauke Reuter for indicating this source.
6. As outlined in the preceding section.
7. The *homo oeconomicus*, a fictive actor is assumed to explain macro-economic phenomena and acts in a self-interested, rational manner to maximise personal benefits with a fixed set of preferences and perfect information.
8. See also the *Journal of Artificial Societies and Social Simulation* (http://jasss.soc.surrey.ac.uk, 4.4.2008).
9. See also the list of papers at http://topics.scirus.com/Agent Based Spatial Models for Agriculture Land Use and Water Management (accessed 11.9.2008)
10. See also similar work by Bousquet and LePage (2004), D'Aquino et al. (2003), and others listed on www.cirad.fr.
11. See also www.gefcoral.org.
12. Marco Janssen, presentation at Stockholm Resilience Conference, April 2008

13. The models by Lansing and colleagues on the long-term social-ecological dynamics surrounding land use and coral reef development on the Indonesian island of Bali (Lansing 2006; Wilson et al. 2007) are truly interdisciplinary in terms of including natural and social dynamics but do not use ground-truthing with model agents
14. Participatory rural appraisal and related fields.

REFERENCES

Anderies, J (2002) The transition from local to global dynamics: A proposed framework for agent-based thinking in social-ecological systems. In: Janssen, MA (ed) *Complexity and Ecosystem Management: The Theory and Practice of Multi-Agent Systems*. Edward Elgar Publishing, Cheltenham, pp 13–34

Anderies, JM, Walker, BH, Kinzig AP (2006) Fifteen weddings and a funeral: Case studies and resilience-based management. Ecology and Society 11: Art. 21 (online)

Arnstein, SR (1969). A ladder of citizen participation. Journal of the American Institute of Planners 35: 215–224

Axelrod, R (1984) *The Evolution of Cooperation*. Basic Books, New York

Barreteau, O (2003) Our Companion modelling approach. Journal of Artificial Societies and Social Simulation 6 (online)

Berger, T, Schreinemachers, P (2006) Creating agents and landscapes for multi-agent systems from random samples. Ecology and Society 11: 19

Berkes, F, Folke, C (1998) *Linking Social and Ecological Systems for Resilience and Sustainability*. Cambridge University Press, Cambridge

Berkes, F, Folke, C (2002) Back to the future: Ecosystem dynamics and local knowledge. In: Gunderson, LH, Holling, CS (eds) *Panarchy: Understanding Transformations in Human and Natural Systems*. Island Press, Washington DC, pp 121–146

Biermann, F (2007) Earth system governance as a cross-cutting theme of global change research. Global Environmental Change 17: 326–337

Billari, FC, Fent, T, Prskawetz, A, Scheffran, J (2006) *Agent-Based Computational Modelling: Applications in Demography, Social, Economic and Environmental Sciences*. Physica Verlag, Springer, Heidelberg, New York

Bousquet, F, Barreteau, O, D'Aquino, P, Etienne, M, Boissau, S, Aubert, S, Le Page, C, Babin, D, Castella, JC (2002) Multi-agent Systems and Role Games: Collective Learning processes for Ecosystem Management. In: Janssen, MA (ed) *Complexity and Ecosystem Management: The Theory and Practice of Multi-Agent Systems*. Edward Elgar, Cheltenham, pp 248–285

Breckling, B, Müller, F, Reuter, H, Hölker, F, Fränzle, O, (2005) Emergent properties in individual-based ecological models-introducing case studies in an ecosystem research context. Ecological Modelling 186: 376–388

Capra, F (2007) Complexity and life. Systems Research and Behavioral Science Wiley Interscience 24: 475–479

Cash, DSW, Adger, N, Berkes, F, Garden, P, Lebel, L, Olsson, P, Pritchard, L, Young, O (2006) Scale and cross-scale dynamics: Governance and information in a multi-level world. Guest Editorial Ecology and Society 11: Art. 8 (online)

Castella, JC, Tran Ngoc Trung, Boissau, S (2005) Participatory simulation of land-use changes in the northern mountains of Vietnam: The combined use of an agent-based model, a role-playing game, and a geographic information system. Ecology and Society 10: Art. 27 (online)

Chambers, R (1992) Rural appraisal: Rapid, relaxed and participatory. Institute of Development Studies, Discussion Paper 311. University of Sussex, Brighton, UK

Cleland, D (2007) Practice makes perfect? An iterative approach to participatory modelling for coral reef management. Submitted for the BSc with Honours in Fenner School of Environment and Society. Australian National University
Defila, R, Di Giulio, A (1998) Interdisziplinarität und Disziplinarität. In: Olbertz, JH (1998) *Zwischen den Fächernßüber den Dingen? Universalisierung versus Spezialisierung akademischer Bildung.* Leske und Budrich, Opladen, pp 111–137
De Vries, B (2006) Scenarios: Guidance for an uncertain and complex world? In: Costanza, R, Graumlich, L, Steffen, W (eds) *Sustainability or Collapse? An Integrated History and Future of People on Earth (IHOPE)*—96th Dahlem Workshop. MIT Press, Cambridge, MA, pp 378–398
Emmeche, C, Køppe, S, Stjernfelt, F (1997) Explaining emergence: Towards an ontology of levels. Journal For General Philosophy of Science 28: 83–119
Epstein, JM, Axtell, R (1996) *Growing Artificial Societies: Social Science From the Bottom Up.* MIT Press/Brookings Institution, Cambridge, MA
Ferrand, N (2007) Modelling with and for stakeholders. In: Phan, D, Amblard, F (eds) *Agent-Based Modelling and Simulation in the Social and Human Sciences.* The Bardwell Press, Oxford, pp 195–217
Flitner, M, Görg, C (2008) Politik im globalen Wandel räumliche Maßstäbe und Knoten der Macht. In: Brunnengräber, A, Burchardt, HJ, Görg, C (eds) *Mit mehr Ebenen zu mehr Gestaltung? Multi-Level Governance in der transnationalen Sozial und Umweltpolitik.* Nomos, Baden-Baden, pp 163–181
Folke, C, Pritchard, L, Jr, Berkes, F, Colding, J, Svedin U (1998) The Problem of Fit between Ecosystems and Institutions. IHDP Working Paper No 2.
Gebetsroither, E, Kaufmann, A, Gigler, U, Resetarits, A (2006) Agent-based modelling of self-organisation processes to support adaptive forest management. In: Billari, FC, Fent, T, Prskawetz, A, Scheffrahn, J (eds) *Agent-Based Computational Modelling Applications in Demography, Social, Economic and Environmental Sciences.* Physica-Verlag, Heidelberg, New York, pp 140–152
Gilbert, N, Troitzsch, K (2005) *Simulation for the Social Scientist.* Open University Press. McGrawHill, Maidenhead, Berkshire
Glaeser, B, Bruckmeier, K, Glaser, M, Krause, G (2009) Social-ecological systems analysis in coastal and marine areas: A path toward integration of interdisciplinary knowledge. In: Lopes, PFL, Begossi, A (eds) *Current Trends in Human Ecology.* Cambridge University Press, Cambridge, pp 180–200
Glaser, M (2003) Ecosystem, local economy and social sustainability: A case study of Caeté estuary, North Brazil. Wetlands Ecology and Management. Special Issue Amazonian Mangroves 11: 265–272
Glaser, M (2006) The social dimension in ecosystem management: Strengths and weaknesses of human–nature mind maps. Human Ecology Review 13: 122–142
Glaser, M, Krause, G, Oliveira, RS, Herazo-Fontalvo, M (2010) Mangroves and people: A social-ecological system. In: Saint Paul, U, Schneider, H (eds) *Mangroves dynamics and Management in North Brazil.* Springer, Heidelberg, Berlin, pp 307–354
Govan, H (2009) Chapter 6: Approaches for Coastal Planning and Resource Management. People, Participation and the Tropical Scientist. In: Wolff, M (ed) *Tropical Waters and Their Living Resources: Their Ecology, Assessment and Management.* H.M. Hauschild, Bremen, pp 172–196
Grimm, V, Berger, U, Bastiansen, F, Eliassen, S., Ginot, V, Giske, J, Goss-Custard, J, Grand, T, Heinz, S, Huse, G, Huth, A, Jepsen, JU, Jørgensen, C, Mooij, WM, Müller, B, Pe'er, G, Piou, C, Railsback, SF, Robbins, AM, Robbins, MM, Rossmanith, E, Rüger, N, Strand, E, Souissi, S, Stillman, RA, Vabø, R, Visser, U,

DeAngelis, DL (2006) A standard protocol for describing individual-based and agent-based models. Ecological Modelling 198: 115–126

Grimm V, Railsback, S (2006) Agent-based models in ecology: Patterns and alternative theories of adaptive behaviour. In: Billari, FC, Fent, T, Prskawetz, A, Scheffran, J (2006) *Agent-Based Computational Modelling: Applications in Demography, Social, Economic and Environmental Science.* Physica, Heidelberg, pp 139–152

Grimm, V, Revilla, E, Berger, U, Jeltsch, F, Mooij, WM, Railsback, SF, Thulke, HH, Weiner, J, Wiegand, T, DeAngelis, DL (2005) Pattern-oriented modelling of agent-based complex systems: Lessons from ecology. Science 310: 987–991

Gunderson, LH, Holling, CS (2002) *Panarchy: Understanding Transformations in Human and Natural Systems.* Island Press, Washington, DC

Gurung, TR, Bousquet, F, Trébuil, G (2006) Companion modelling, conflict resolution, and institution-building: Sharing irrigation water in the Lingemuteeychu Watershed, Bhutan. Ecology and Society 11: Art. 36 (online)

Halliday, A, Glaser, M (2011) A management perspective on social ecologicalsystems: A generic system model and its application to a case study from Peru. Human Ecology Review 18:1–18

Hare, M, Letcher, RA, Jakeman, AJ (2003) Participatory modelling in natural resource management: A Comparison of four case studies. Integrated Assessment 4: 62–72

Holland, JH (1998) *Emergence: From Chaos to Order.* Helix Books, Reading

Holling, CS, Gunderson, LH (2002) Resilience and adaptive cycles. In: Gunderson, LH, Holling, CS (eds) *Panarchy: Understanding Transformations in Human and Natural Systems.* Island Press, Washington, DC, pp 25–62

Holling, CS, Gunderson, LH, Ludwig, DH (2002) In quest of a theory of adaptive change. In: Gunderson, LH, Holling, CS (eds) *Panarchy: Understanding Transformations in Human and Natural Systems.* Island Press, Washington, DC, pp 3–22

Huigen, M, Overmars, KP, de Groot, W (2006) Multiactor modelling of setting decisions and behavior in the San Mariano Watershed, the Philippines: A first application with the MameLuke framework. Ecology and Society 11: Art. 33 (online)

Jager, W, Janssen, MA (2001) Diffusion processes in demographic transitions: A prospect on using multi-agent simulation to explore the role of cognitive strategies and social interactions. SOM research report 01B40

Jager, W, Janssen, MA (2002) Using artificial agents to understand laboratory experiments of common-pool resources with real agents. In: Janssen, MA (ed) *Complexity and Ecosystem Management: The Theory and Practice of Multi-Agent Systems.* Edward Elgar Publishers, Cheltenham, Northhampton, pp 75–102

Jager, W, Janssen, MA, De Vries, HJM, De Greef, J, Vlek, CAJ (2000) Behaviour in commons dilemmas: Homo economicus and Homo psychologicus in an ecological-economic model. Ecological Economics 35: 357–379

Janssen, MA (ed) (2002) *Complexity and Ecosystem Management: The Theory and Practice of Multi-Agent Systems. Edited Papers from the Biannual Meeting of the International Society for Ecological Economics.* Edward Elgar, Cheltenham, Northhampton

Janssen, MA, Alessa, LN, Barton, M, Bergin, S, Lee, A (2008) Towards a community framework for agent-based modelling. Journal of Artificial Societies and Social Simulation 11 (online)

Janssen, MA, Ostrom, E (2006) Empirically based, agent-based models. Ecology and Society 11: Art. 37 (online)

Johannes, RE (1981) *Words of the Lagoon: Fishing and Marine Lore in the Palau District of Micronesia.* University of California Press, Berkeley

Kroeze, C, Middelburg, J, Leemans, R, Fennel, W, Glaser, M, Harshima, A, Liu, KK, Meybeck, M (2008) Integrating Tools to assess change in semi-enclosed marine systems. In: Urban ER, Sundby, B, Malanotte-Rizzoli, P, Melillo, J (eds) *Watersheds, Bays and Bounded Seas. The Science and Management of Semi-Enclosed Marine Systems*. Island Press, Washington, DC, pp 77–96

Lansing, JS, Kremer, JN (1993) Emergent properties of balinese water temple networks: Coadaptation on a rugged fitness landscape. American Anthropologist 95: 97–114

Lipton, M (1978) *Assessing Village Labour Situations in Developing Countries: Study Prepared for the International Labour Office*. Institute of Development Studies, Oxford

Manson, SM (2002) Validation and verification of multi-agent models for ecosystem management. In: Janssen, MA (ed) *Complexity and Ecosystem Management: The Theory and Practice of Multi-Agent Systems. Edited Papers from the Biannual Meeting of the International Society for Ecological Economics*. Edward Elgar, Cheltenham, Northhampton, pp 63–74

Nicolson, CR, Starfield, AM, Kofinas, GP, Kruse, JA (2002) Ten heuristics for interdisciplinary modelling projects. Ecosystems 5: 376–384

Olsson, P, Folke, C, Hahn, T (2004) Social-ecological transformation for ecosystem management: The development of adaptive co-management of a wetland landscape in southern Sweden. Ecology and Society 9: Art. 2 (online)

Ommer, RE (2007) *Coasts under Stress: Restructuring and Social-Ecological Health*. McGill-Queens University Press, Montreal

Pereira, A, Duarte, P, Reis, LP (2004) *Agent-Based Simulation of Ecological Models. Proceedings of 5th Workshop on Agent-Based Simulation*. SCS Europe, BVA

Pyka, A, Grebel, T (2006) Agent-based modelling: A methodology for the analysis of qualitative development processes. In: Billari, FC, Fent, T, Prskawetz, A, Scheffran, J (eds) *Agent-Based Computational Modelling: Applications in Demography, Social, Economic and Environmental Sciences*. Physica, Heidelberg, pp 17–35

Reuter, H, Hölker, F, Middelhoff, U, Jopp, F, Eschenbach, C, Breckling, B (2005) The concepts of emergent and collective properties in individual-based models: Summary and outlook of the Bornhöved case studies. Ecological Modelling 186: 489–501

Rudd, M, Railsback, S, Danylchuk, A, Clerveaux, W (2002) Developing a spatially explicit agent-base model of queen conch distribution in a Marine Protected Area in the Turks and Caicos Islands. In: *Proceedings of the Gulf Caribbean Fisheries Institute*. GFCI, Marathon, pp 259–271

Sawyer, K (2005) *Social Emergence: Societies as Complex Systems*. Cambridge University Press, Cambridge

Shafer, JL (2007) Agent-based simulation of a recreational coral reef fishery: Linking ecological and social dynamics. PhD Dissertation at the University of Hawaii, Department of Geography

Spencer, L, Ritchie, J, Dillon, L (2003) *Quality in Qualitative Evaluation: A Framework for Assessing Research Evidence*. Government Chief Social Researcher's Office

Uchmanski, J, Grimm, V (1996) Individual-based modelling in ecology: What makes the difference? Trends in Ecology and Evolution 11: 437–441

Wilson, J, Yan, L, Wilson, C (2007) The precursors of governance in the Maine lobster fishery PNAS 1004: 15212–15217

Yan, L, Wilson, J, Wilson, C (2008) Modelling the precursors of governance in the Maine lobster fishery. In: Chapman, RL (ed) *Creating Sustainability Within Our Midst: Challenges for the 21st Century*. Pace University Press, New York pp 85–98

9 Integrated Modelling and Scenario Building for the Nicobar Islands in the Aftermath of the Tsunami

Martin Wildenberg and Simron Jit Singh

INTRODUCTION

Many of our global and local sustainability problems can only be understood and tackled by abolishing the dualistic view that separates the natural from the social realms. Consequently, there is increasing interest in the notion of the coupled social-ecological system (SES). Generally speaking, an SES can be conceptualised by considering human society as the arena of interaction between the natural and cultural spheres of causation and as being composed of these three elements, that is, man, nature and culture (Fischer-Kowalski and Weisz 1999; see also Tretter and Halliday, this volume). Practitioners of sustainability science generally conceptualise SES in similar ways, but a number of different operational approaches have been developed. Two influential approaches are (1) social metabolism and (2) the resilience concept.

The 'socio-metabolic approach' places emphasis on the biophysical relationship between society and nature and is based on the premise that a society organises material and energy exchanges with its environment to satisfy its requirements for reproduction and maintenance. The quantity and structure of matter and energy a society draws from its environment largely depends on people's mode of subsistence and lifestyle, which is in turn related to technology. The higher the throughput of matter and energy per capita, the greater is the impact on the environment. There is increasing resource scarcity on the input side and, on the output side, decreasing capacity of the ecosystem to continue absorbing wastes and emissions (Ayres and Simonis 1994; Fischer-Kowalski 1998; Haberl 1999; Haberl 2001).

The 'resilience approach', on the other hand, conceptualises SES as complex adaptive systems, thereby laying emphasis on their dynamic behaviour and the inherent uncertainties therein. Within this framework, the two central concepts proposed are 'resilience' and the 'adaptive cycle'. Resilience refers to the behaviour of the system in response to external disturbances and internal changes; the adaptive cycle describes four phases commonly observed in many natural, social and coupled systems in which the system shows different degrees of resilience. Adaptive management is recommended for dealing with complex adaptive SES. Management interventions may be seen as real-life experiments that contribute to a better understanding of the system and therefore to better management (Folke et al. 2002; Holling et al. 2002; Carpenter et al. 2001; Walker et al. 2006; Olsson et al. 2004).

Within sustainability science there have been several attempts to model SES at different scales from global to local. The input–output model approach is often used to characterise the metabolic profile of social-ecological regimes in terms of their material and energetic throughput (e.g., Fischer-Kowalski and Haberl 1997; Fischer-Kowalski and Haberl 2007; Weisz 2007; Sangwon 2005). For example, global vegetation models like the Lund-Potsdam-Jena (LPJ) Dynamic Global Vegetation Model (www.pik-potsdam.de/~kirsten/web_page_lpj.htm) provide insights into the effects on vegetation of climate change and increasing atmospheric CO_2 concentrations (Sitch et al. 2003). Other models simulate land use and land cover change caused by human and natural drivers (Parker et al. 2003). Models of local SES that simulate the interactions between the different actors and their environment have been used for scenario building or conflict resolution (Bousquet and Le Page 2004). The resilience approach employs a range of dynamic simulation models (Janssen and Carpenter 1999; Holling 1973), both for theoretical purposes and as a decision support tool. Scenario development can provide inputs to participatory processes aimed at formulating policies for local resource use (Barreteau et al. 2003; Walker et al. 2002). In a recent article, Epstein (2008) underscores the variety of possible uses of models and names sixteen reasons for modelling "other than prediction". He states that the main advantage of a simulation model is that it provides an explicit statement of what otherwise is only implicitly contained in qualitative mind maps.

In this chapter, we present preliminary results of a model of an SES of Kamorta Island (henceforth referred to as the K-SES model) belonging to the Nicobar archipelago in the Bay of Bengal. It is one of the very few attempts to link input–output metabolic modelling with an agent-based land-use change model rooted in the complex adaptive system approach. The modelled system includes the island of Kamorta with its human population and the surrounding reefs and shallow waters. The Nicobar Islands were subject to immense destruction by the 2004 tsunami, and now face choices between different possible pathways in the rehabilitation process. The model is seen as a means to evaluate the various possible scenarios. It is also envisioned that such a model could be used in a participatory process to inform stakeholders of the likely medium-term consequences of their decisions. In this paper, we present the underlying concepts and design of the island SES model and preliminary results. These results provide a mathematically expression of linkages between different components of the SES. We quantify the material and energy exchanges between the terrestrial and marine resource systems, conceived of as supply systems (Hummel, this volume), and social systems using inputs from demographic analysis and information about household preferences in production and consumption. The version of the model used for the simulation runs presented here focuses on the terrestrial part of the SES and provides the basis for a more comprehensive application. As some modules, such as the economic sub-module or the reef module, still have to be calibrated and tested, the results presented here are only fragmentary and are intended to give a foretaste of what is to come.

DESCRIPTION OF THE STUDY AREA: NICOBAR ISLANDS

The Nicobar archipelago is a group of 24 islands in the Bay of Bengal, administered as an Indian union territory. These islands are home to a rich tropical biodiversity with several endemic terrestrial and marine species and inhabited by an indigenous community, commonly referred to as the Nicobarese. Mongoloid in origin and having migrated from the Malay-Burma coast over 2,000 years ago, the Nicobarese have remained relatively isolated for a long time. However, owing to their geographical location on an important sea route, the islands were often visited by passing vessels with the aim of replenishing food and water supplies on the long and arduous sea voyages of colonial times. Consequently, a small amount of barter trade took place, where the Nicobarese exchanged food and coconuts for cloth and iron, and later rice, tobacco and other consumables from time to time. The British colonised the islands in 1869 and for the first time regulated trade and set up an administrative system. In 1947, the islands became part of independent India. Since 1956 the islands have been protected and access has been regulated under the Andaman and Nicobar Protection of Aboriginal Tribes Regulation (ANPATR). Today access to the islands is still limited to Indian citizens, who also require a permit (Singh 2003).

The Island of Kamorta covers 188 km^2 and is the largest of the central Nicobar Islands. The predominant land cover of the island is made up of different types of evergreen and semi-evergreen forest (about 40% of the total area) and *Imperata* grasslands (about 35%); remaining areas are mangroves and shrubland (Gupta et al. 2004). *Imperata cylindrica* is known throughout Southeast Asia as an aggressive, highly adaptive weed, able to thrive in poor soils and difficult to eradicate because of its tough rhizomes that rapidly invade agricultural fallows (Chikoye 2003). Grasslands are burned regularly and used as hunting grounds for wild boar (Singh 2003).

The central Nicobars are rich in biodiversity with high rates of endemism especially among birds. Kamorta, which harbours 31 of the 35 endemic bird taxa found on the Central Nicobars, is the most important island from a conservation point of view (Sankaran 1997). The coast and tidal creaks are lined with extensive stretches of mangrove forest (covering about 650 ha) and in the waters around Kamorta relatively pristine coral reefs (about 1,600 ha) can be found, which are exploited only for subsistence purposes. Over 135 coral species, and some 2,000 fish species, have been recorded in the shallow waters of the Andaman and Nicobar Islands (Hoon 1997). Due to the action of ocean currents, the marine diversity of the Andaman and Nicobar Islands has greater affinity to Southeast Asia than to other reef areas in the South Asian Region. Threats to the marine system arise from plans to develop this 'underexploited' resource and from increasing pressure from poaching of sea urchins and shark fins by fishermen from Burma and Thailand (State Development Report of Andaman & Nicobar Islands 2006).

The Nicobarese have a largely subsistence economy. They make a living from hunting, gathering, fishing, coconut production and rearing pigs and chickens. Some families practice horticulture and have home gardens where they grow fruit, vegetables and a variety of roots and tubers. Coral reefs are an important source of protein-rich seafood. The link to the market is via the production and sale of copra (dehydrated coconuts) in exchange for rice, sugar, cloth, fuel, toiletries and other consumables. Thus, coconuts compose an important source of livelihood, both on a subsistence level (a third of the coconut production is fed to pigs) and as a commercial product in the form of copra. Coconut plantations are frequently planted, as this act secures the ownership of the land, but subsequent maintenance is poor. Coconuts are also harvested from naturally occurring coconut palms.

The tsunami of December 2004 turned the world of the Nicobarese literally upside down and initiated a 'complex disaster' on the islands (see Singh 2009). Owing to their close proximity to the epicentre of the earthquake, the low lying islands and their inhabitants were subject to immense devastation. In a matter of minutes, some 4,000 people were swept away by the gigantic waves. Their villages were either completely destroyed or damaged beyond repair. Losses included not only property and livestock but also cultural artefacts—some of them hundreds of years old. All coconut trees along the coast within a kilometre from the sea were washed away or killed as the seawater passed over. Losses of anchored boats led to a breakdown of communication between villages and the administrative headquarters. The earthquake and the consequent sinking of the island land masses resulted not only in the destruction of more than half of the mangrove forests and about 40% of the coral reefs but also in the loss of large areas of land to the ocean, creating a new coastline and making navigation more difficult (Singh 2006; Ramachandran et al. 2005).

The national and international response to the tsunami which ravaged Southeast Asia was overwhelming. Approximately 10 billion dollars were donated or pledged in what has been called the world's largest fund-raising exercise. The Nicobars were one of the worst affected areas, and there was a race among the NGOs to get access to the islands, which represented an opportunity to expand their development activities. Although they were determined to be self-reliant in the first few months after the tsunami, the Nicobarese were gradually overwhelmed with aid of all kinds: housing, free food, enormous amounts of cash compensation and a variety of consumables and material goods. Such an inflow of aid in cash and kind did not come without consequences. Conflict arose over the distribution of resources, leading to a breakdown of the joint family structure and property rights, increasing social stratification between the haves and have-nots, and changes in the political sphere. A good leader was one who brought in more aid, and in increasing amounts. The benevolence of the government in compensating losses with cash was a predicament in itself. Never before had the Nicobarese had so much money at one time. Before the tsunami, whenever money was required to buy something, copra was made and immediately sold. Capital accumulation among the Nicobarese was rare, if not unknown; the concepts of investment and saving for future needs were also unknown[1]. Consequently, the cash received from aid agencies was spent immediately and there was an increase in the purchase of consumer goods such as motor-bikes, TVs, DVD players, mobile phones, music systems and processed food.

The new economic order being promoted by the Indian government will have direct and indirect effects on the development path of the islands. The short-term aim of these measures was to compensate for lost property and ensure the continuity of supplies to satisfy people's basic needs. The longer term goal is to reduce the vulnerability of the local population to natural hazards and to diversify the economy, reducing dependency on subsidies from the Indian government and bringing the islands up to 'international standards' (Andaman & Nicobar Administration 2006). Measures taken in this context have included the rebuilding of destroyed villages further away from the coast, the promotion of vegetable and spice gardens for subsistence and cash crop production and the promotion of village cooperatives to market fish or other local products. New infrastructure like roads had to be built to connect the now landlocked villages (Andaman & Nicobar Administration 2008).

Aid programmes from the Indian government and international aid organizations have affected traditional social and power relations, leading to an erosion of traditional institutions, values and customary rules for resource use. Aid money has accelerated the transition from a subsistence based economy towards an economy linked in to the global market and dependent on aid money and imported goods. Such a process is usually accompanied by changes in the relations between a society and its environment both at a cultural and biophysical level, and in this respect the Nicobar Islands are no exception.

MODELLING THE SOCIAL-ECOLOGICAL SYSTEM OF KAMORTA ISLAND

Methodological Challenges

SES and other complex adaptive systems are characterised by a large number of interacting components, non-linear and emergent behaviour, multi-scale interactions, the ability to self-organise, multiple stable states and path-dependent development (Levin 1998). Models of SES incorporate humans and human actions as well as non-human components of these systems. They pose particular difficulties due to the complex and diverse motivations and processes driving human behaviour. However, the goal of modelling is not to reproduce the entire SES in all its complexity, but to provide answers to specific questions of interest to the modellers. A model needs to be complex enough to answer the question which is being asked—but not too complex. It should selectively incorporate relevant attributes of the SES and its dynamics that provide inputs to answering research questions, but at the same time maintain the holistic vision necessary to address wider sustainability related questions (Wu and Marceau 2002). Another important challenge of SES modelling is to validate the outcomes of the model. Modellers have proposed various validation approaches, but their usefulness is the subject of continued debate (e.g. Moss 2008; Küppers et al. 2005). In the case of the K-SES model, empirical model testing is difficult due to the novelty of the situation, as well as ongoing structural changes in the system. Therefore, an evidence-driven approach (Moss 2008) was chosen, drawing on existing field data and expert knowledge. Another form of validation would be to test the usefulness of the model with stakeholders (Sterman 2002). However constraints often arise from limited availability of data, which might lead to a certain level of reductionism for practical reasons.

Our own model belongs to the class of simulation models. The purpose of a simulation model can either be to predict how the system might behave in the future under assumed conditions (such as changes in policy, market behaviour, etc.) or to appraise policy proposals by testing their effect on the system (Sterman 1991). Our own model had both these aims. We wanted to understand how the system might behave and look like in future as a consequence of aid interventions and government policies in the aftermath of the tsunami. We also hoped the model would serve as a tool to inform consideration by stakeholders of possible future policy options.

For these purposes, we needed software that would help to meet both objectives and allows the combination of different modelling paradigms, specifically input–output and agent-based scenario building approaches. We selected AnyLogic (URL 2), a Java-based software which also provides templates for designing user-friendly interfaces.

Nevertheless, while the choice of appropriate software can help in providing the necessary tools, it is only a first step towards overcoming the challenges of dealing with complex adaptive SES. In what follows, we list the challenges we faced in modelling the Kamorta SES and briefly describe our attempts to address each of them.

Selection of Appropriate Components and Processes

Modelling the SES of Kamorta Island required the inclusion of a distinct set of elements from both natural and social systems. Here the notion of social metabolism was useful in helping us to focus on the exchange of materials and energy between society and nature. We followed these flows and explicitly modelled them together with their endogenous and exogenous drivers. As our starting point, we took population, lifestyle and consumption patterns as the main driver of demand for subsistence and consumption goods. We then looked at the various options the Nicobarese had for organising their production system to meet this demand. The main source of livelihood was found to be the extraction and transformation of terrestrial and marine natural resources. A small percentage of the population obtained cash incomes by providing services and wage labour. These considerations led us to select the following components for inclusion in the model: human population, household demand (influenced by consumption patterns and lifestyle), systems of production, land use, and relevant terrestrial and marine ecosystem dynamics that interact with the systems of production. The detail which had to be put into the different sub-components was evaluated with reference to the questions to be analysed by the model.

Dealing with Multiple Scales

All living systems are organised in a nested, hierarchical way. We attempted to capture this reality in our K-SES as well. First, we gathered information on social and natural processes operating at the various hierarchical levels of the system, from individual households to the whole island. We used different tools to generate data on the interactions between these scales and processes. Secondly, we incorporated the different scales of the system into the model, by matching the different components to the scales at which they were differentiated. Thus lifestyle preference and consumption are modelled at the household scale; vegetation and land use at the landscape (terrestrial ecosystem) scale and tsunami aid and markets for products at the global (whole island) scale.

Problem of Relevant Time Scales

As well as spatial resolution, it is also important to consider different time scales. Calculating time requirements and available working time on a yearly basis for instance may hide seasonal peaks in working time that are typical of agricultural systems, which can act as constraints on the sizes of agricultural holding that can be maintained in production. To address this issue the K-SES, in addition to operating on yearly time steps, also estimates maximum monthly working hours during the year by considering labour requirements during the two peak agricultural seasons[2].

Addressing Emergent Behaviour and Shift Between States

Emergent behaviour is one of the key features of SES and is linked to cross-scale interaction. In our model the individual behaviour of agents in the social system and changes in land use lead to the emergence of patterns at the landscape level such as aggregations of fields, plantations or settlements. Cross-scale interactions between the ecosystem and landscape level also determine the shift between alternative land-cover states. For example, the extent of the grassland determines the probability of fire, which in turn affects the development towards either forest or grassland states on the landscape scale.

Path Dependent Behaviour

SES are characterised by path-dependent development. That is, the evolution of the system depends on the historic legacy of the path it has so far taken. In the model, path-dependent behaviour is introduced where relevant in the different components, with the result that path dependency is also observable on the system level. On the ecosystem level, path dependency results from transition rules which depend on past states of the ecosystem, which mimic successional states or the building up of stocks through time. For example, biomass accumulation leads to increasing fire risk. To calculate future agricultural productivity of an area we consider data from previous years to arrive at an average annual yield per unit area. To take another example, we assume that an agent's choice of land-use for a particular area of land would be dependent on this area's land-use history.

Modelling Human Behaviour

Human behaviour has been approached in several ways in the modelling community, for example, based on rational choice theory, limited rationality, (economic) utility functions or heuristics (Berger and Parker 2002; Janssen 2004). In the last decade, agent-based modelling has increasingly been applied for modelling land use (spatially explicit models) and its interactions with the social system. Agents, in this context refer to individual humans or collective entities such as institutions, associations, unions (Weiss 1999). Since our model is a spatially explicit one linking social and ecological systems we decided to use an agent based approach. This approach provides flexibility in dealing with scales, supports the integration of different data types, and makes it easier to represent landscapes, nature and societies as interacting entities as they are commonly perceived. In the current configuration, the whole social system (i.e., the collective Nicobarese population of Kamorta Island) is represented by one agent only. Decisions taken by this agent obey rules which allow it to respond to its perception of certain aspects of its environment and its own (internal) condition. Some decisions, which were considered more complex or where we lacked empirical data to determine appropriate algorithms, were modelled using sliders. These allow the critical parameters determining the behaviour of the agent to be changed during model run-time in an interactive process. This unique feature allows stakeholders to interact or play with the model. They can create realistic scenarios and learn about the response of the system under different conditions. Additionally this feature can be used to incorporate new information collected about stakeholder preferences and patterns of behaviour into the model.

Dealing with Complex Systems in Transition

The dynamics, uncertainties and instabilities of the post-tsunami situation in the Nicobar Islands (and Kamorta Island in particular) led to frequent changes in the focus and aim of the model. This required a flexible model and an iterative modelling process. For optimal results, the features of the computer model (structure and components) should complement the organization of the modelling process (Richardson 2002). To achieve the necessary flexibility, the computer model was designed in a modular way. Using a modular design means that elements can be modified, added or removed easily if required without influencing the performance of the remaining modules. Being able to run the single modules as stand-alone models also makes model testing and calibration much easier. It is also possible to use the K-SES model at different scales; either to simulate the whole island system or (after some adaptations) to simulate single plots.

Data Sources and Dealing with Knowledge Gaps

There is limited published data or literature about the island. For the pre-tsunami period, we were fortunate to be able to draw on the five years of (published and unpublished) research results of Simron Singh on society–nature interactions, including data on both biophysical and cultural aspects (Singh et al. 2001; Singh and Grünbühel 2003, Singh 2003, 2006). Data from remote sensing and aerial photos were crucial for initialising the K-SES model. We also used information published by the Indian Institute of Remote Sensing (Roy et al. 2003) that includes maps showing different land-cover types. For the post-tsunami period, aerial photos from Google Earth (accessed in 2008) were used. These show newly built settlements and roads, and also some of the destruction caused by the tsunami. Maps of coral reefs were available from Pacific ReefGIS on ReefBase (pacificgis.

reefbase.org/). Additional post-tsunami data was specially collected for the model and related to demography, household consumption, working time preferences, production patterns, land-use preferences, and the availability and use of terrestrial and marine resources. For this various methods were used, such as standard interviews, observation, interviews assisted by the use of small Excel models, spider-grams (Lynam 2002) and focus groups (Kasemir 2003). Thus, the model combines traditional and scientific knowledge, an approach which has often been emphasised as a prerequisite for managing SES in sustainability literature (e.g., Gadgil et al. 2003). While a more direct involvement of the local population in the model development and design would have been desirable, this was not possible for practical reasons. Additionally, secondary literature was used where available.

During the modelling process we were able to identify crucial data and knowledge gaps. Some of these gaps could be filled with little effort, while others required innovative thinking. For example, some of the knowledge gaps were addressed using 'pattern oriented modelling' (Grimm and Railsback 2005). The behaviour of any single living system depends on a range of site and context specific parameters including legacies from the past (Ulanowicz 1997: 38). At the same time, we have amassed a wealth of information on general system behaviour in the last decades. Combining context specific information together with general rules on the behaviour of systems as we know them helped us to crosscheck and test our data and provided additional insights into system behaviour. In this way the model allowed us to estimate unknown parameter values by comparing the patterns emerging in the model with real-world observations documented in the literature. Despite these attempts to fill knowledge gaps during the modelling process, some data could not obtained in time to be incorporated into the model. For example, spatially explicit elevation data was not available and thus we could not tackle the question of erosion. However, since the model in its current state is still adaptable to new data inputs, missing data can be easily incorporated into subsequent versions. There is an easy-to-use data interface that links the model to GIS systems.

Model Structure

The model consists of four interlinked modules: (1) socio-economic, (2) agent-based, (3) ecosystem and (4) output (see Figure 9.1). The socio-economic module interacts with the ecosystem module via the agent that simulates human decisions. The combined effect of these decisions are then exhibited or displayed over the output module as social and environmental indicators. The model has a user-friendly interface that allows the stakeholders to engage in an interactive process and play around with their decisions and test their assumptions on system behaviour. The following sections describe the four modules in more detail.

The Socio-Economic Module

The social system of the inhabitants of Kamorta Island is modelled in a separate module using various modelling approaches, and as such may form an independent unit. This module contains three sub-modules, namely, demography, household lifestyle and the economy. The demographic component simulates changes in population structure over the next 20 years and was developed by the International Institute for Applied Systems Analysis (IIASA). The output of this sub-module is age, class and gender specific population data. Households are classified into three lifestyles: traditional, moderate and modern. The speed (and direction) of transition from one household type to another can be set to simulate the transition from a subsistence based to a more monetised society. The output of this sub-module is the population per household type, which is then used to calculate the demand for money (to buy desired consumer goods) and, together with outputs from the demographic sub-module, the total available working time. The central component of the economic sub-module is a Leontief input–output matrix, which is used to calculate the total demand for money and resources (including tools and materials necessary in the production process). The results indicate the island's total requirement for natural resources and other materials, in order to meet household demand for money and

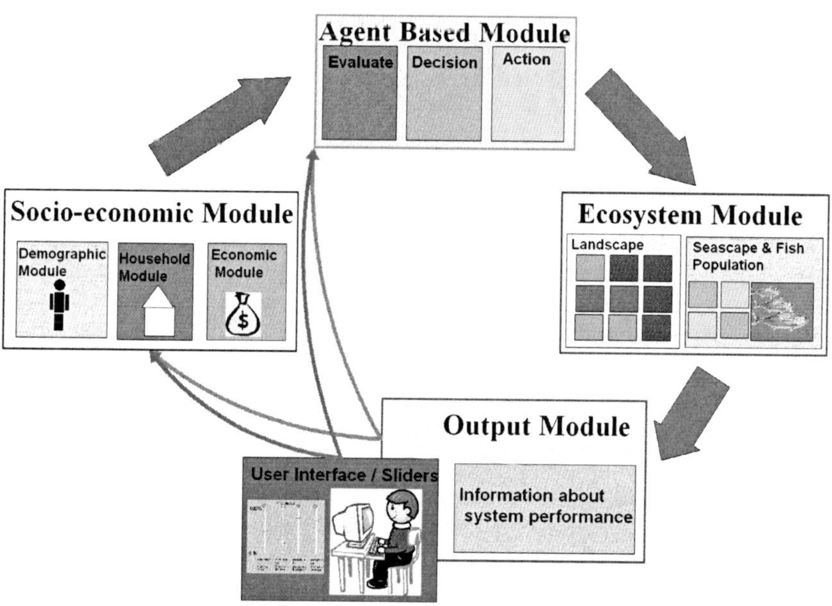

Figure 9.1 Schematic diagram of the 'K-SES-Model'.

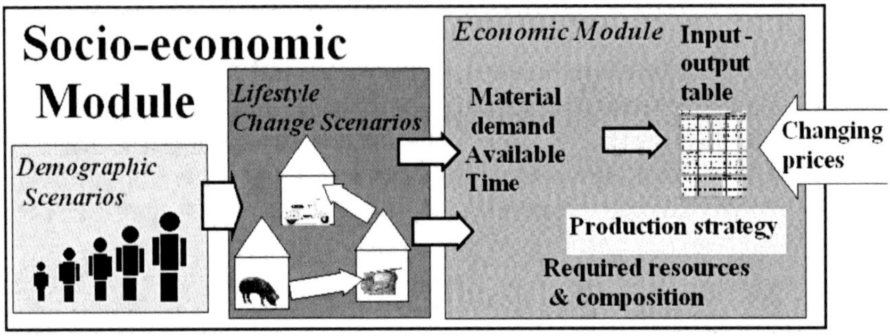

Figure 9.2 Schematic diagram of the 'Socio-Economic Module'.

subsistence products under a given scenario. The matrix also calculates the amount of working time needed to meet these requirements.

The Agent-Based Module

The agent-based module receives as an input from the society module the total amount of natural resources that need to be harvested for the reproduction of the population with a given distribution of lifestyle types. The basic decision an agent has to take in the K-SES model is whether it wants to maintain, increase or decrease the amount of land under cultivation. To reach this decision the harvest of the previous year is compared to the current demand for resources. The area under cultivation is increased or decreased so that the expected harvest matches current demand. In this case, the agent does not have a choice to intensify land use and increase productivity per unit area of land, since we assume that the Nicobarese will not be able to switch to another technology in the foreseeable future. Factors which the model considers when calculating the required area of land include declining yields due to loss of fertility in home gardens, stochastic fluctuations in yields, and changes in the yields of coconut plantations over time.

The agent takes decisions about the use of land unit cells (termed LANDUNITs) in the ecological model. The agent perceives two variables of a LANDUNIT that influence his decision. One indicates its suitability for a certain land-use type and the other the efficiency of the current land use in terms of yield or money per unit of time. The suitability indicator, which is represented as a number between 0 and 8, is calculated taking account of the current land cover, land use in neighbouring areas, and the land-use history of the LANDUNIT. Through changing the weights, the Suitability Index can be adjusted to the preferences (and perception) of the actors. Ownership, which is at the moment only considered in the small-plot version of the model, or other factors at the LANDUNIT level, could be introduced to provide additional inputs to land-use decisions. To select LANDUNITs (e.g., to transform into a garden), the agent randomly roams through them and chooses those that fulfil the predetermined suitability criteria (e.g., when the Suitability Index for a garden ≥ 6.5). When all LANDUNITs have been considered a further iteration can be performed, lowering the criteria until the requirements are met. This mimics a situation where the agents do not have complete information about the conditions of every single LANDUNIT but rather investigate each piece of land for its usability and decide to use it if it fulfils the minimum criteria—regardless of whether there might be a better piece of land somewhere else.

If excessive production occurs, the least efficient LANDUNITs are abandoned first. Efficiency can be either calculated as money earned per hour of work invested or in biophysical units such as material or energy gained per hour of work. Required working time is contingent on how accessible the selected LANDUNITs are and their land-use types. In contrast to the method used for choosing LANDUNITs (where we assume that the agent does not have complete knowledge), in the case of abandonment we assume that the agent does know about the efficiency of all the LANDUNITs he is maintaining and will first abandon the least efficient.

Also a 'minimum efficiency value' can be set which will lead to the agent abandoning each LANDUNIT that exhibits efficiency below this value. As both suitability and efficiency of a LANDUNIT are influenced positively by neighbouring colonised LANDUNITs, patterns of aggregation emerge on the landscape level. Assigning different weights to neighbourhood relations can produce different patterns of land use on the landscape level ranging from aggregated to dispersed patterns of land use. The proportion of different products to be sold in the market (i.e., consumer demand) determines what is produced (coconut, vegetables, fruits and fish) and also preferences between forest and grassland locations, and whether to settle close to or far from the coast. These preferences can be changed during the model run via the user interface.

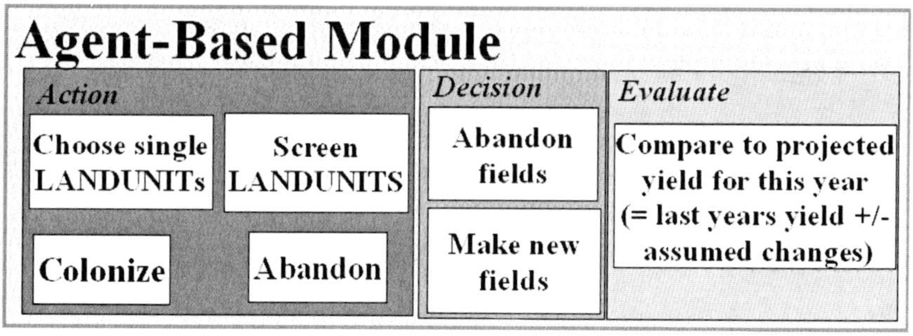

Figure 9.3 Schematic diagram of the 'Agent-Based Module'.

The Ecosystem Module

The ecosystem module consists of two sub-modules, terrestrial (or landscape) and marine, which together simulate ecosystem dynamics. In this chapter, however, we present results from the terrestrial part only. The terrestrial sub-module consists of a grid of LANDUNIT cells. Each LANDUNIT has x- and y-coordinates, which define its exact position and a number of parameters and variables describing its land cover, its land-cover change history, and the probability of fire. A rule based 'state chart' controls the dynamics of the LANDUNITs and the changes between different land-cover types. In our model we use 15 land-cover and land-use states, of which 4 are colonised states: vegetable gardens, fruit gardens (orchards), coconut plantations and settlements. LANDUNITs can shift from one land-cover state to another either through internal time-driven processes which represent succession, through fire, or through a colonisation process by humans initiated through the agent-based module (Figure 9.4).

Most of these transitions are influenced to some extent by the land-cover states of the neighbouring LANDUNITs. Fire outbreaks occur stochastically in the model. The probability of occurrence is determined by land-cover state and the proximity to roads or to other human influenced LANDUNITs. The probability of fire spreading from one LANDUNIT to a neighbouring one depends on its land-cover type and the time since the last fire (representing the fuel load).

Yields and working time required to maintain the colonised LANDUNITs, that is, of vegetable gardens, coconut plantations and orchards are dependent on the internal condition of each LANDUNIT (such as age of crops and trees, factors affecting soil, and proximity to human settlement). The time required for all economic and subsistence activities on each LANDUNIT is calculated from field data. We also accounted for other

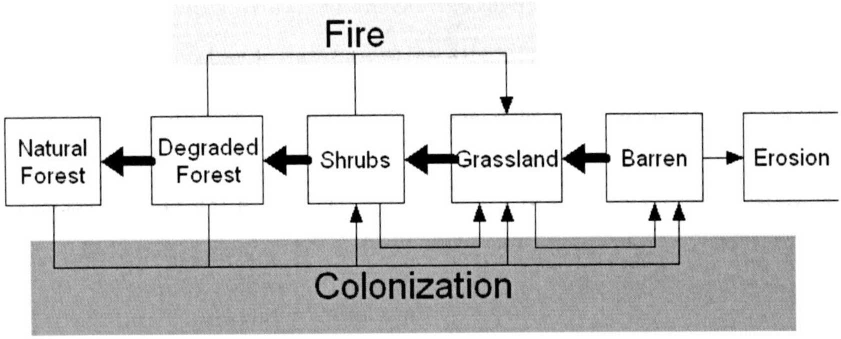

Figure 9.4 Schematic representation of the state-chart for land-units. The thick arrows represent the succession process through time. The thin arrows show state changes either induced by fire or through the process of abandonment after colonisation.

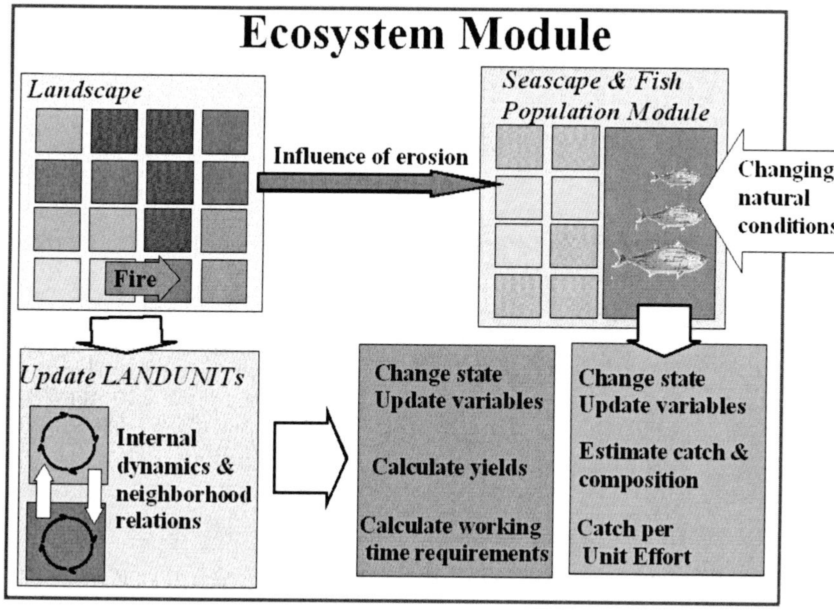

Figure 9.5 Schematic diagram of the 'Ecosystem Module'.

factors affecting time use such as accessibility and neighbourhood conditions. Yields can be expressed in money, kilograms and energy units per hectare and per unit of time. The use of both energy and money as currencies in the model makes it possible to link subsistence, monetary and mixed economic systems.

The Output Module

The output module displays the combined effects of human decisions and changes in framework conditions on the various socioeconomic and ecosystem variables, thereby providing information on system performance. The results calculated by this module are automatically fed back to the computer agent and influence the agent's decisions and actions in accordance with a set of pre-defined rules derived from field observations. However, the model also provides a user interface that makes it possible for real agents (stakeholders) to interact with the model. With decreasing soil fertility as a determining factor, they may directly intervene in the model by playing around with a set of sliders representing various system parameters and human preferences. In other words, the Nicobarese have the possibility to react to the output results and adjust their preferences (e.g., consumption patterns, land-use choices) represented by sliders and optimise the desired outcomes. Presently, we have introduced a set of five (land-use preference) sliders and check boxes for the users to influence agent behaviour during

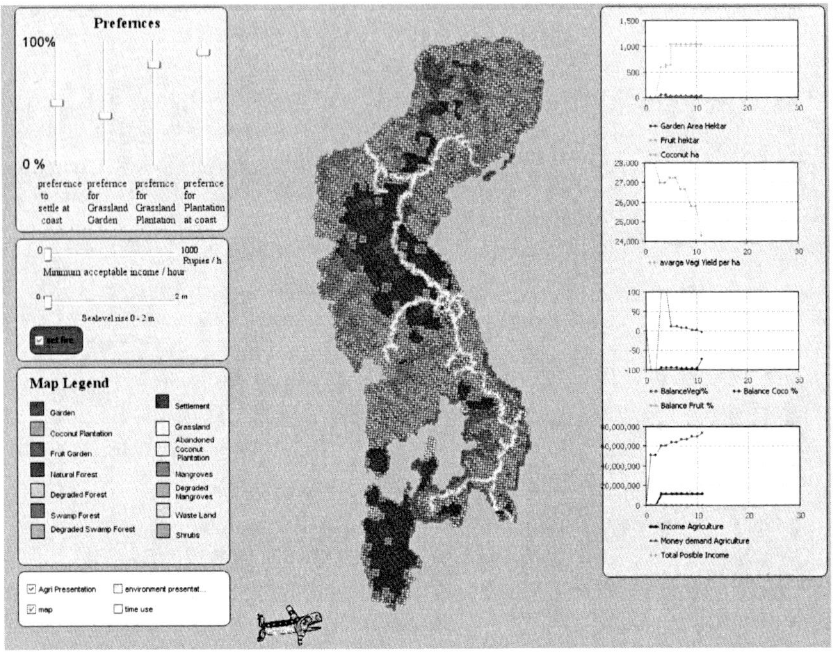

Figure 9.6 Screenshot of the model user interface with sliders in the top left corner, map of Kamorta and output graphs on the right.

run-time of the model. The model surface displays a map of Kamorta Island showing land use and land cover, which are affected by human decisions. Thus, the aggregated outcome of human decisions is manifested in changes in land use and land cover on the map of Kamorta Island. Additionally, the surface displays outputs of some relevant variables such as income and yield per unit of land or hour, loss of forest to agriculture, and working time required to meet the needs of the population with a given mix of lifestyles. This allows the agent or the user to specify constraints such as maximum available time or to compare activities like gardening with fishing in terms of monetary returns or yield per hour of work invested.

Baseline Scenario and Assumptions

The model takes the year 2005 as its starting point, a year after the tsunami, and the baseline data used in the model has been regularly updated (to 2008). In 2005, Kamorta Island had 1,741 inhabitants and 440 households, with an adult working population (across gender) of 1,095 in the age group 15–60. The average overall population growth rate is taken to be 1.5% per year, resulting in a projected increase in the number of households from 440 to 740 households over 30 years[3].

As discussed already, the humanitarian aid and several welfare programmes initiated by the NGOs and the Indian government have triggered changes in the lifestyle and consumption patterns of the islands. Many modern amenities and commodities have been introduced, and there is ongoing construction of new infrastructure. Obviously, these new patterns of consumption cannot be met by a subsistence economy alone. Besides the reestablishment of coconut plantations (since most of the old ones were destroyed in the tsunami), one of the major interventions by the Department of Agriculture of the Indian government has been to promote the introduction of agriculture in the form of commercial fruit and vegetable production. The new economy based on agriculture is expected to generate substantial income that will enable the Nicobarese to meet their needs.

We use the model to simulate two scenarios—one based on slow and the other on rapid modernization—and investigate the associated impacts on land cover and land use, and on working time and incomes. We assume that household consumption is the main driver of demand for resources and cash incomes, and that this demand must be met by agriculture. We used interviews to estimate the consumption levels of individual households and categorised them as low, moderate or high consumption types, each representing a distinct pattern of consumption. The low ("traditional") consumption category represents a lifestyle and consumption patterns close to what they were before the tsunami, with a low demand for cash mainly for the purchase of rice, sugar, tea and consumables like toiletries, clothes, and fossil fuels for boats. Moderate consumption households, in addition, purchase a certain amount of packaged snacks, soft drinks, and inexpensive electronic products like watches or radios. High consumption ("modern") households possess relatively expensive electronic goods like TV sets, DVD players, refrigerators, mobile phones, and motorbikes, in addition to consuming packaged snacks and drinks. In the model, the different consumption patterns are reflected by the different monthly cash requirements of the three household types: 8,291 rupees (€130), 11,186 rupees (€175) and 13,479 rupees (€211), respectively[4]. Field research found a striking similarity between household types and their willingness to work. Low consumption households expressed their willingness to do approximately one hour's work per day to generate cash incomes;[5] for moderate consumption households the figure was 2.5 hours, and for high consumption households up to 4 hours' work per person per day.

The transition of households from low to moderate to high consumption profiles in the model represents the process of modernisation. While change is an inherent characteristic of all SES, change processes were accelerated in the Nicobars by interventions after the tsunami. As mentioned previously, we model two scenarios of change: slow and rapid. Field observations indicated that a plausible 'rapid modernisation' process would be characterised by the number of modern households growing at an annual average rate of 5% while number of the moderate and traditional households would fall at an annual rate of 7% and 3%, respectively. A 'slow modernisation' process would represent a predominant shift to a moderate lifestyle rather than a modern one. In this scenario, both numbers of modern and moderate households grow at an average of 2% per year (but the number of moderates households grows more in absolute terms) while numbers of traditional households fall by 0.7% per year.

Between 2005 and 2008, the Department of Agriculture established coconut plantations on 110 hectares of reclaimed land and 482 hectares of fresh grassland on Kamorta Island[6]. However, the coconut palms will begin to fruit only after seven years, with productivity reaching a maximum by 2015. In addition, 28 hectares of grassland were converted to vegetable gardens; crops include ocra, cowpea, gourds, eggplant, chilli and leafy vegetables. Since intercropping of the different vegetables is promoted, the model currently assumes a single garden type. The Department of Agriculture estimates vegetable yields of between 66 and 81 tons per hectare per year, which is very close to the Indian average in the FAO database (FAOStat 2005). This, however, is a best-case scenario that could be met only under optimal conditions including fertilisation, and may not be applicable to Nicobar conditions. The Nicobarese are only just learning to grow vegetables and much lower annual yields have been recorded in at least one village, ranging between 6 and 8 tons per hectare. Constraints that contribute to this low yield include the lack of irrigation facilities, soil conditions, and lack of experience and unwillingness to work on the part of the Nicobarese. To be more realistic, therefore, our model assumes a range of 25 to 30 tons per hectare per year. In the model, yields start declining after the field has reached a certain age, to represent the degradation of soil quality that is typical for many tropical soils. As no information on the soil quality of Kamorta exists, we assumed values as found in other tropical regions and made it easy to manipulate these settings to test the consequences of different parameter values in the model.

As mentioned already, an important output in the model is the working time required for subsistence as well as for commercial production. Subsistence work is the time required to obtain fish, coconuts and vegetables for household consumption. Production for the market takes into account additional time required for processing and marketing for copra or vegetables. At present, we consider fishing to be mainly a subsistence activity for household consumption only with only some fish sold at the local markets. The time required to create new coconut plantations or vegetable gardens depends on the original land cover, with forest requiring considerably more time (8,500 hours per hectare) than grassland (1,280 hours per hectare). The model also takes into account market conditions and assumes that local markets can absorb a maximum of 365 tons of vegetables per year at an average price of 30 rupees per kilogram[7]. Sales of larger amounts would depend on getting access to export markets.

RESULTS AND DISCUSSION

In its current configuration, the socio-economic module incorporates a simplified version of the economic sub-module; the ecosystem module is also simplified and represents only terrestrial systems. Household demand for resources and money over the next 30 years is seen as the main driver of change and has consequences on land cover and land use as well as on working time required to meet the total demand. We begin with the latter, and then go on to discuss the environmental impacts (in terms of land use and land cover) using the two scenarios—slow and rapid modernisation.

Demand on Working Time to Meet Household Consumption

The results show that the investment of working time required over the next years varies considerably. In an initial period there is a steep increase in the required working time due to the need to establish and maintain new coconut plantations and vegetable gardens. With a start date of 2005, this process is completed by 2008 (see Figure 9.7). The following years are fairly stable. The time required for work on the vegetable gardens and for fishing is about three hours per working adult per day. From 2015, when the coconut palms are mature, additional time for copra production is required. From this time on, until the end of the model run, the working time requirement per adult is more then eight hours per day: the entire disposable working time.

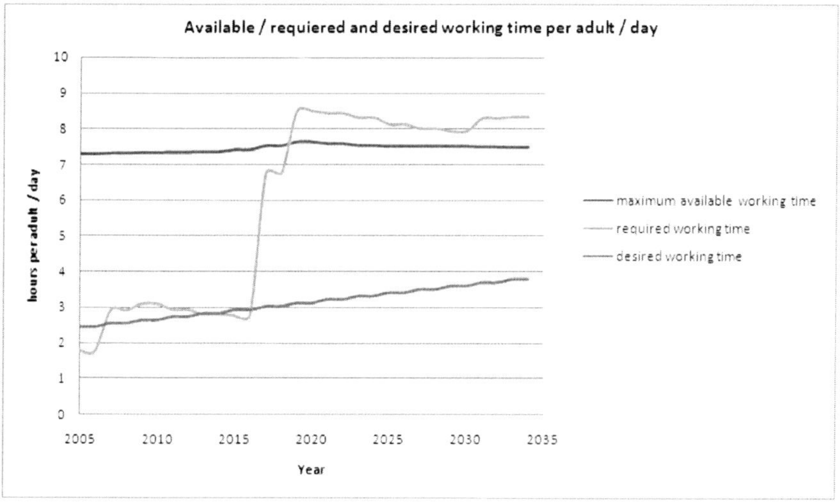

Figure 9.7 Available, required and desired working times per adult. Required working time includes time for subsistence and market oriented activities.

During the entire period it is assumed that the Nicobarese actually produce and sell all the vegetables and fish that the local market can potentially absorb (i.e., 1,000 kg of vegetables and 500 kg of fish each day). Thus, on average each household needs to maintain only 0.06 hectares of vegetable garden, assuming an annual yield of 25–30 tons per hectare. In principle the Nicobarese would have enough time to produce more fish and vegetables for the market, but in the absence of export facilities (transportation and storage infrastructure, access to markets), the extra production would not produce any benefits. In this sense, even with the best scenario of selling all the fish and vegetables that they can, the Nicobarese can meet only 40% of their household income per month—that is 3,390 of the estimated 8,250 rupees required per household, for the rest they would have to rely on aid and cash reserves. It is only after 2015, when copra can be produced and exported again, that the Nicobarese are able to meet the remaining 60% of their household needs, albeit with a high investment of working time.

Is there a way out of this dilemma? In other words, can income be increased with low time inputs? Figure 9.8 shows that it is copra production that absorbs most time; much more than fishing and vegetable or fruit production. However, copra production produces less income per hour of work. An hour of vegetable production generates an income of 55 rupees, fishing generates 20 rupees per hour (assuming current average costs and prices on the local market) while copra brings in only 13 per rupees per hour. It would thus seem to make sense to increase vegetable respectively fruit production and fish catch, and for export facilities to be provided to

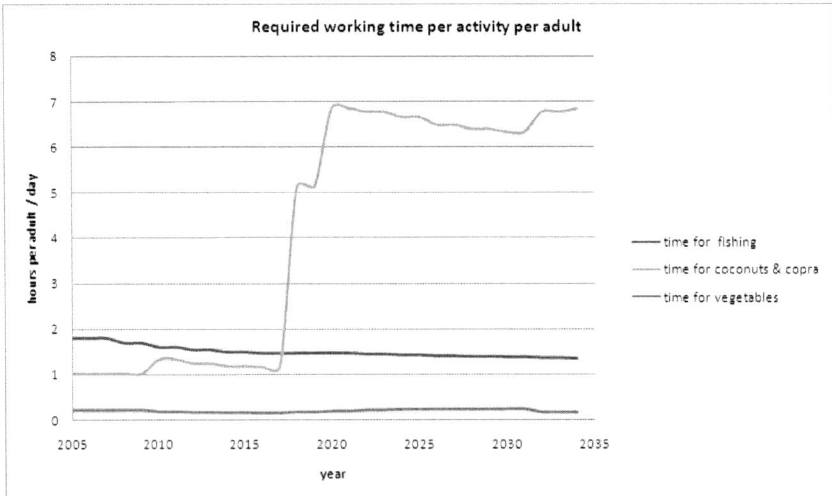

Figure 9.8 Working time required per adult to meet consumption demands, split up into time required for fishing, vegetable and coconut gardening and processing.

enable produce to be sold to nearby islands and beyond, in the same way as the government has provided facilities and subsidies for copra trade. In other words, if the Nicobarese wish to continue their present consumption levels without becoming dependent on free food and aid, efforts should be made to expand the export potential of vegetables and fish.

Changes in Land Use and Land Cover and Related Impacts

The results from the model show major changes in land use and the land cover of the island over the next 30 years. In order to fulfil the current demand for money and subsistence produce the Nicobarese require 5 hectares per household of coconut plantations; that is, a total area of about 3,500 to 4,000 hectares on Kamorta Island. Annual yields are assumed to be about 7 tons per hectare, slightly lower than the Indian average[8]. Due to the limited capacity of the local market to absorb vegetables, the total area of vegetable gardens required is only between 26 and 55 hectares. Although it is envisaged that most of the coconut plantations would be established on grasslands (80% in the model), this still implies a total loss of forest of about 2,500 hectares over the next 30 years. This loss is due not only to direct conversion of forest into plantations or gardens but also to degeneration leading to a state shift from a forest to a grassland-dominated system. Degeneration occurs as a result of fuel wood extraction for copra production, estimated to be about 4,500 tons dry weight per year, which puts a great deal of pressure on the forest. Interestingly, the scale of these impacts does not differ greatly between the two scenarios. In the case of rapid modernisation, an additional 300 hectares are required for coconut plantations, of which 25 hectares are currently forest.

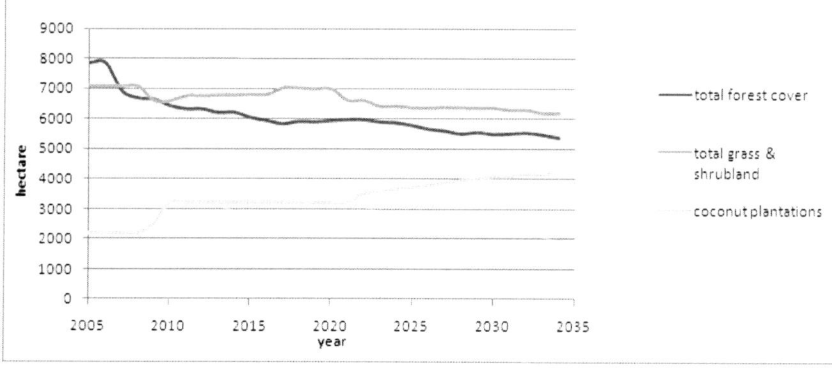

Figure 9.9 Change in forest, grassland and shrub land and coconut plantations area in the next 30 years.

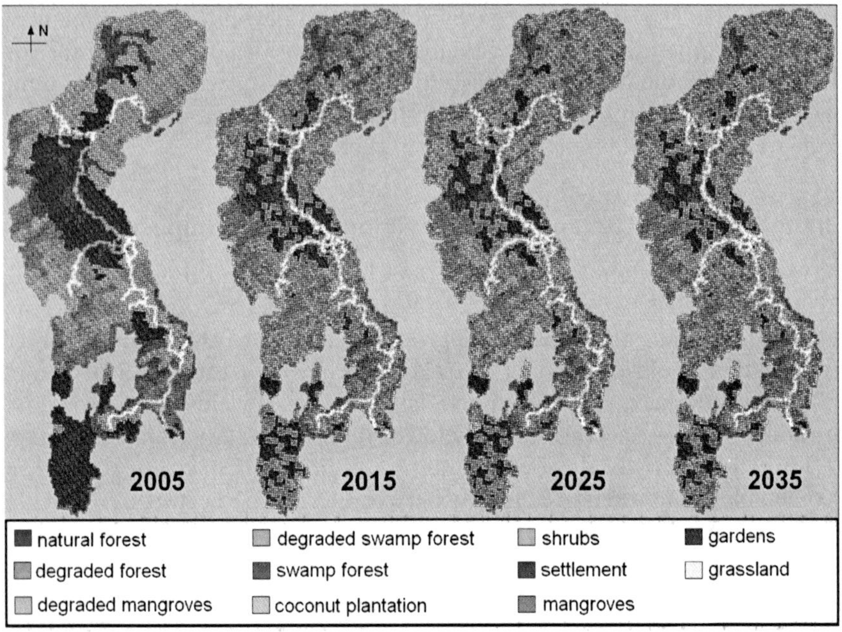

Figure 9.10 Kamorta land cover and land use: 2005, 2015, 2025 and 2035.

Besides direct and indirect forest loss, there is also an increased fragmentation of the forest whereby the length of the forest border increases in relation to the total forest area. The ratio between border length to forest area increases from 0.18 to 0.47. Figure 9.9 shows the changes in areas of grassland, forest, coconut plantations and vegetable gardens over 30 years.

Figure 9.10 shows the land use and land cover of Kamorta Island in 10-year intervals, starting from 2005 to 2035. The model result indicates that the two large forest patches existing in the year 2005 become increasingly fragmented over the next 30 years. The central forest is reduced in size and split into three parts.

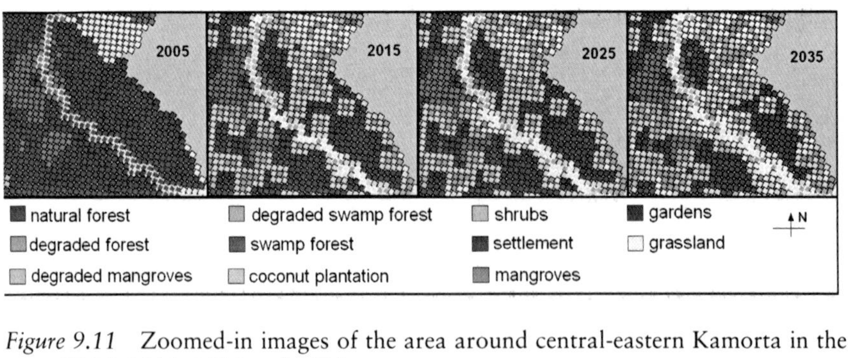

Figure 9.11 Zoomed-in images of the area around central-eastern Kamorta in the years 2005, 2015, 2025 and 2035.

If we zoom in (Figure 9.11), we can see how forest fragmentation occurs due to intrusion of grassland into the forests along the road in the central eastern part of the island, as well as due to 'edge effects' on the borders between forest and colonised land. These patterns are similar to those found in other SES with similar environmental and economic conditions (e.g., Tomich et al. 1996).

The rate of fragmentation also depends on the preferences of the agents in selecting new land for cultivation. As vegetable growing is new on these islands the Nicobarese have still to find out which types of land are most suitable. One of the measures taken by the Department of Agriculture to reduce pressure on the forest has been to establish most new coconut plantations on grasslands, in mixed plantations that enable inter-cropping with other products such as cashew nut. However, the Nicobarese fear that once their plantations or gardens are turned into a 'forest', the forest department will claim that land and restrict its future use.

The combined effects of land-use changes due to economic activities are not without consequences for fragile ecosystems on small islands. Most of the problems encountered are connected to water shortages and soil erosion. Loss of forest cover and land-use conversion are the main drivers of this process on most tropical islands (Millennium Ecosystem Assessment 2003). Additional ecosystem services provided by forests on small islands include shore line protection, provision of food—including famine food in critical times—medicinal plants, fuel, wood for construction and biodiversity conservation (Wilkie et al. 2002). The extent of forest loss predicted for Kamorta may have a negative impact on drinking water quality and quantity, on soil erosion and soil fertility, on the availability of forest products and on the conservation status of some of the endemic fauna and flora found on the island. The decline in water availability, combined with a likely increase in the demand for water due to population growth and changes in lifestyle, is a potentially serious problem. If climate change predictions for the region are accurate, changes in rainfall patterns could also have a negative influence on the hydrological system of the islands. Moreover, under these conditions, rain-fed agriculture would become more difficult, and increased reliance on irrigation would place additional pressure on this scarce resource. Last but not least, soil erosion might also have negative effects on the coral reefs surrounding the islands (e.g., Carpenter et al. 2008).

CONCLUSIONS

The tsunami of 2004 has opened up new opportunities for the Nicobarese, but at the same time it confronts them with new challenges and uncertainties about their future. While the tsunami in itself was devastating, the aid interventions did little to bring the islands to a new stable state. We have seen that the impact of the interventions not only affects the natural system of the island but also creates a huge discrepancy between the newly adopted lifestyle and the local capacity to satisfy the islanders' needs. We do not know how the Nicobarese would react if the aid flow ceased; whether they would revert to consumption levels as they were before the tsunami or would gradually adapt to working longer hours. The latter scenario seems rather improbable. It may very well be that new export opportunities for fish and vegetables are opened up to the islanders, perhaps with additional possibilities for the export of value-added products such as processed fish and spices. This would increase their income per hour and allow them to meet their increasing demand for money with a modest increase in working time and without undue pressure on land and other natural resources.

Modelling efforts such as the one described in this chapter provide valuable insights into the dynamics of complex systems and throw light on the range of opportunities and bottlenecks that may affect the outcome of policy options being considered for the islands. In an effort to build scenarios for the Nicobar Islands, we made progress in the field of modelling by being able to link the agent-based approach with that of input-output modelling. Further, the use of time as an important variable allowed us to understand the social consequences of changing lifestyles and land-use practices.

The future of the Nicobarese will depend on their capability to deal with all the surprises, uncertainties and opportunities which the 2004 tsunami washed onto their shores. The outcome of current interventions on the island will only gradually become clear and it is in the interest of all stakeholders, especially the Tribal Council and the Indian Administration, that their results are monitored to identify successes, failures and unexpected side-effects, so that appropriate remedial action can be taken. This would be in line with the principles of learning-based adaptive management. Modelling can contribute to learning and adaptation by helping stakeholders to understand and reflect on the intricate relationships between various system components that give rise to often unexpected outcomes in the real world.

ACKNOWLEDGMENTS

The authors are grateful to the Austrian Science Foundation (FWF) for its financial support in developing the model (Project RECOVER, L 275-G05). The second author is particularly thankful to the Andaman and Nicobar Administration for logistical support during fieldwork, and to the RAI Fellowship in Urgent Anthropology (2009–2011) that supported the writing of the results. We are particularly grateful to our colleagues Marina Fischer-Kowalski, Willibald Haas, Helmut Haberl, Veronika Gaube and Heidi Adensam for their valuable time in discussions and comments. We appreciate the support of colleagues from the International Institute for Applied Systems Analysis (IIASA)—Wolfgang Lutz, Warren Sanderson, and Samir K.C.—in discussions and help with the demographic analysis.

NOTES

1. Traditionally the Nicobarese did accumulate capital in the form of pigs which were an important status symbol and were needed in sufficient amounts for certain festivals and rites. It is questionable though that this form of "pig accumulation" can be easily compared to the world of money-based economy (Wildenberg 2005).
2. In contrast to usual sociological time use research, we are mainly interested in human time as a resource on the social system level for a sustainability analysis. For analytical reasons, we distinguish between four relevant functional subsystems of the social system that are reproduced by a certain activity: the person system, the household system, the economic system and the community system (for details, see Singh et al. 2010). Thus, working time here refers to the time invested in reproducing the economic system both for subsistence as well as for the market. Nicobarese, however, will consider rituals and ceremonies as work since they serve the function of pleasing the spirits for continued abundance of natural resources, and so are inextricably linked to the economic system. However, for analytical reasons and to allow for comparisons across cases in time-use studies within sociology and anthropology, we prefer to take the etic view.
3. The population growth scenario taken for our purpose assumes an increasing life expectancy at birth while the Total Fertility Rate (TFR) at the start is set to 2 and decreases over time. This seems to be the most probably scenario given the success of several government medical and birth control programmes.
4. The cost of durables such as TVs, DVD players, refrigerators and motorbikes is spread over the period they are expected to last under local conditions (climate, handling, etc.).
5. Pre-tsunami time-use studies have revealed that the Nicobarese actually spent no more than an hour a day in economic activities (Singh et al. 2001).
6. The agriculture department encouraged inter-cropping of one cashew nut tree in between four coconut palms to assist in diversifying the economy of the Nicobarese. In the model, however, we account only for the area under coconut plantations.
7. The consumers in the local market comprises of government officials, armed forces personnel and mainland settlers that in part are working under contract for the government in the construction sector; in total about 5,000 people.
8. According to the agriculture department, the Indian average is 6,863 nuts per hectare and the average weight of a coconut is 1.2 kilograms (Wenkam 1990). However, the yields per hectare in the Nicobars are reported to be much lower (3,698 nuts or 4.4 tons per hectare) due to high density of coconut palms and their poor maintenance. Still, in our model we assume a higher productivity, assuming that the new plantations would be supervised by the Department of Agriculture and would be expected to give comparable yields.

REFERENCES

Andaman & Nicobar Administration Announcement (2006) URL: http://www.and.nic.in/Announcements/post-tsunami.pdf (accessed October 2008)

Andaman & Nicobar Administration Reports (2008) URL: http://tsunamiandaman.tn.nic.in/ (accessed October 2008)

Ayres, RU, Simonis, UE (1994) *Industrial Metabolism: Restructuring for Sustainable Development.* United Nations University Press, Tokyo, New York, Paris

Barreteau, O, Antona, M, D'Aquino, P, Aubert, S, Boissau, S, Bousquet, F, Daré, W, Etienne, M, Le Page, C, Mathevet, R, Trébuil, G, Weber, J (Collectif Commod) (2003) Our companion modelling approach (La modélisation comme outil d'accompagnement). Journal of Artificial Societies and Social Simulation 6 (online)

Berger, T, Parker, DC (2002) Examples of specific research. In: Parker, DC, Berger, T, Manson, SM (eds) *Agent-Based Models of Land-Use and Land-Cover Change: Report and Review of an International Workshop, Irvine, California,* LUCC Report Series No. 6 pp. 79—87

Bousquet, F, Le Page, C (2004) Multi-agent simulations and ecosystem management: A review. Ecological Modelling 176: 313–332

Carpenter, KE, Abrar, M, Aeby, G, Aronson, RB, Banks, S, Bruckner, A, Chiriboga, A, Cortés, J, Delbeek, JC, DeVantier, L, Edgar, GJ, Edwards, AJ, Fenner, D, Guzmán, HM, Hoeksema, BW, Hodgson, G, Johan, O, Licuanan, WY, Livingstone, SR, Lovell, ER, Moore, JA, Obura, DO, Ochavillo, D, Polidoro, BA, Precht, WF, Quibilan, MC, Reboton, C, Richards, ZT, Rogers, AD, Sanciangco, J, Sheppard, A, Sheppard, C, Smith, J, Stuart, S, Turak, E, Veron, JEN, Wallace, C, Weil, E, Wood, E (2008) One-third of reef-building corals face elevated extinction risk from climate change and local impacts. Science 321: 560–563

Carpenter, S, Walker, B, Anderies, JM, Abel, N (2001) From metaphor to measurement: Resilience of what to what? Ecosystems 4: 765–781

T.J. Terry (1994) Imperata cylindrica (L.) RAEUSCHEL In: Labrada, R, J. C. Caseley (ed) *Weed Management for Developing Countries Addendum 1.* Food And Agriculture Organisation Of The United Nations (FAO), Rome pp61—69

Epstein, JM (2008) Why model? Journal of Artificial Societies and Social Simulation 11 (online)

Fischer-Kowalski, M, Weisz, H (1999) Society as hybrid between material and symbolic realms: Toward a theoretical framework of society–nature interrelation. Advances in Human Ecology 8: 215–251

Fischer-Kowalski, M (1998) Society's metabolism. The intellectual history of material flow analysis, Part I: 1860–1970. Journal of Industrial Ecology 2: 61–78

Fischer-Kowalski, M, Haberl, H (1997) Tons, joules, and money: Modes of production and their sustainability problems. Society and Natural Resources 10: 61–85

Fischer-Kowalski, M, Haberl, H (2007) *Socioecological Transitions and Global Change: Trajectories of Social Metabolism and Land Use.* Edward Elgar, Cheltenham, UK

Folke C, Carpenter, S, Elmqvist, T, Gunderson, L, Holling, C (2002) Resilience and sustainable development: Building adaptive capacity in a world of transformations. AMBIO: A Journal of the Human Environment 31: 437–440

Gadgil, M, Olsson, P, Berkes, F, Folke, C (2002) Exploring the role of local ecological knowledge in ecosystem management: three case studies. In: Berkes, F, Colding, J, Folke, C (eds) *Navigating Social-Ecological Systems—Building Resilience for Complexity and Change.* Cambridge University Press, Cambridge pp. 187—189

Grimm, V, Railsback, SF (2005) *Individual-Based Modeling and Ecology.* Princeton Series in Theoretical and Computational Biology. Princeton University Press, Princeton, NJ

Gupta S, Porwal, MC, Roy, PS (2004) Human modification of the tropical rain forest of Nicobar islands: Indicators from land use land cover mapping. Journal of Human Ecology 16: 163–171

Haberl H (2001) The energetic metabolism of societies Part I: Accounting concepts. Journal of Industrial Ecology 5: 11–33

Haberl, H, Schandl, H (1999) Indicators of sustainable land use: Concepts for the analysis of society-nature interrelations and implications for sustainable development. Environmental Management and Health 10: 177–190

Holling, CS (1973) Resilience and stability of ecological systems. Annual Review of Ecology and Systematics 4: 1–23

Holling CS, Carpenter, S, Brock, WA, Gunderson, LH (2002) Discoveries for sustainable futures. In: Gunderson, LH, Holling, CS (eds) *Panarchy: Understanding Transformations in Human and Natural Systems*. Island Press, Washington D. C., pp. 395–417

Hoon, V (1997) Coral reefs of India: Review of their extent, condition, research and management status In: Hoon, V, Chong, KC, Roy, R, Bierhuizen, B, Rubens, J, Kanvinde, H (eds) *Regional Workshop on the Conservation and Sustainable Management of Coral Reefs Series Proceedings No .22*. CRSARD, Madras

Janssen MA (2004) Agent based models. In: Proops, J, Safonov, P (ed) *Modelling in Ecological Economics: Current Issues in Ecological Economics*. Edward Elgar, Cheltenham, UK pp. 155—172

Janssen, MA, Carpenter, SR (1999) Managing the resilience of lakes: A multi-agent modeling approach. Conservation Ecology 3: Art. 15 (online)

Kasemir, B, Jäger, J, Jaeger, CC, Gardner, MT (2003) *Public Participation in Sustainability Science: A Handbook*. Cambridge University Press, Cambridge

Küppers, G, Johannes, L (2005) Validation of simulation: Patterns in the social and natural sciences. Journal of Artificial Societies and Social Simulation 8 (online)

Levin, SA (1998) Ecosystems and the biosphere as complex adaptive systems. Ecosystems 1: 431–436

Lynam, T, Bousquet, F, Le Page, C, d´Aquino, P, Barreteau, O, Chinembiri, F, Mombeshora, B (2002) Adapting science to adaptive managers: Spidergramms, belief models, and multi-agent systems modeling. Conservation Ecology 5: Art. 24 (online)

Millennium Ecosystem Assessment (2003) Island systems. Chapter 23 of *Ecosystems and Human Well-Being*. Island Press, Washington D.C.

Moss, S (2008) Alternative approaches to the empirical validation of agent based models. Journal of Artificial Societies and Social Simulation 11 (online)

Olsson P, Folke, C, Berkes, F (2004) Adaptive comanagement for building resilience in social–ecological systems. Environmental Management 34: 75–90

Parker, D, Manso, S, Janssen, MA, Hoffman, M, Deadman, P (2003) Multi-agent systems for the simulation of land-use and land-cover change: A review. Annals of the Association of American Geographers 93: 314–337

Ramachandran, S, Anitha, S, Balamurugan, V, Dharanirajan, K, Ezhil Vendhan, K, Preeti Divien, MI, Senthil Vel, A, Sujjahad Hussain, I, Udayaraj, A (2005) Ecological impact of tsunami on Nicobar Islands (Camorta, Katchal, Nancowry and Trinkat). Current Science 89: 195–200

ReefGIS (2008) provided by UNEP-WCMC URL: http://www.reefbase.org/gis_maps/default.aspx

Richerdson, KA (2002) Methodological implications of complex system approaches to sociality: Some further remarks. Journal for Artificial Societies and Social Simulation 5 (online)

Roy, PS (2003) *Biodiversity Characterisation at Landscape Level in Andaman and Nicobar Islands Using Satellite Remote Sensing and Geographic Information System*. Indian Institute of Remote Sensing & Department of Environment and Forest, Dehradun

Sangwon, S (2005) Theory of materials and energy flow analysis in ecology and economics. Ecological Modelling 189: 3–4

Sankaran, R (1997) Developing a protected area network in the Nicobar islands: The perspective of endemic avifauna. Biodiversity and Conservation 6: 797–815

Singh, SJ (2003) *In the Sea of Influence: A World System Perspective of the Nicobar Islands.* Lund Studies in Human Ecology 6. Lund University Press, Lund

Singh, SJ (2006) *The Nicobar Islands: Cultural Choices in the Aftermath of the Tsunami.* Czernin Verlag, Vienna

Singh, SJ (2009) Complex disasters: The Nicobar Islands in the grip of humanitarian aid. Geographische Rundschau—International Edition 5: 48–56

Singh SJ, Grünbühel, CM (2003) Environmental relations and biophysical transitions: The case of Trinket Islands. Geografiska Annaler, Series B, Human Geography 85: 191–208

Singh SJ, Grünbühel, CM, Schandl, H, Schulz, N (2001) Social metabolism and labour in a local context: Changing environmental relations on Trinket Island. Population and Environment 23: 71–104

Singh, SJ, Ringhofer, L, Haas, W, Krausmann, F, Fischer-Kowalski, M (2010) *Local Studies Manual: A Researcher's Guide for Investigating the Social Metabolism of Rural Systems.* Social Ecology Working Paper 120. IFF Social Ecology, Vienna (available online)

Sitch, S, Smith, B, Prentice, IC, Arneth, A, Bondeau, A, Cramer, W, Kaplan, J, Levis, S, Lucht, W, Sykes, M, Thonicke, K, Venevsky, S (2003) Evaluation of ecosystem dynamics, plant geography and terrestrial carbon cycling in the LPJ Dynamic Vegetation Model. Global Change Biology 9: 161–185

Sivakumar, K (2008) How could a natural catastrophe impact the ecology of a species? The Nicobar megapode and tsunami. Nature Precedings (online)

State Development Report of Andaman & Nicobar Islands (2006) National Institute of Public Finance and policy 18/2, Satsang Vihar Margspecial Institutional Area New Dehli 110 067; URL: http://www.nipfp.org.in/State%20Development%20Report%20of%20A&NIFinal%20(June%2006).pdf

Sterman, JD (1991) A sceptic's guide to computer models. In: Barney, GO, Kreutzer, WB, Garrett, MJ (eds) *Managing a Nation: The Microcomputer Software Catalog.* West View Press, Boulder pp. 209–229

Sterman JD (2002) All models are wrong: Reflections on becoming a systems scientist. System Dynamics Review 18: 501–531

Tomich, TP, Kuusipalo, J, Menz, K, Byron, N (1996) Imperata economics and policy. Agroforestry Systems 36: 233–261

Ulanowicz, RE (1997) *Ecology, the Ascendant Perspective: Complexity in Ecological Systems.* Columbia University Press, New York

Walker, BH, Carpenter, S, Anderies, J, Abel, N, Cumming, GS, Janssen, MA, Lebel, L, Norberg, J, Peterson, GD, Pritchard, R (2002) Resilience management in social-ecological systems: a working hypothesis for a participatory approach. Conservation Ecology 6: Art. 14 (online)

Walker, BH, Gunderson, LH, Kinzig, AP, Folke, C, Carpenter, SR, Schultz, L (2006) A handful of heuristics and some propositions for understanding resilience in social-ecological systems. Ecology and Society 11: Art. 13 (online)

Weiss, G (1999) *Multiagent Systems: A Modern Approach to Distributed Artificial Intelligence.* MIT Press, Cambridge, MA

Weisz, H (2007) Combining social metabolism and input–output analyses to account for ecologically unequal trade. In: Hornborg, A, McNeill, JR, Martinez-Alier, J (eds) *Rethinking Environmental History: World System History and Global Environmental Change.* Rowman & Littlefield, AltaMira pp. 289–306

Wenkam N (1990) *Foods of Hawaii and the Pacific Basin. Research Extension Series, 070.* HITAHR, Honolulu

Wildenberg, M (2005) *Ecology, Rituals and System-Dynamics: An Attempt to Model the Socio-Ecological System of Trinket Island.* IFF Social-Ecology Working Paper Number 80 URL: http://www.uniklu.ac.at/socec/downloads/wp80.pdf

Wilkie, ML, Eckelmann, CM, Laverdiére, M, Mathias, A (2002) *Forests and Forestry in Small Island Developing States*. Forest Management Working Paper FM 22. FAO Forestry Department, Rome
Wu, J, Marceau, D (2002) Modelling complex ecological systems: An introduction. Ecological Modelling 153: 1–6

INTERNET REFERENCES

URL 1:
http://www.illegal-fishing.info/item_single.php?item=news&item_id=2906&approach_id=30
and:
http://www.planetark.org/dailynewsstory.cfm?newsid=36755&newsdate=13-Jun-2006
URL 2:
www.anylogic.com

Part V

Pathways Towards New Systems Thinking in Human Ecology

10 Towards Global Sustainability Analysis in the Anthropocene

Marion Glaser, Gesche Krause, Andrew Halliday and Bernhard Glaeser

INTRODUCTION AND SCOPE

The environmental crises of the 21st century show that humanity has become a planetary force of environmental change and ever more vulnerable to it. New approaches towards analysing how humans and non-human nature interrelate are required. This book puts human–nature relations at the centre of the sustainability debate and explores avenues towards improved dialogue and collaboration between the social and natural sciences.

Humans continuously engage in material and ideational interactions with nature. Technology modifies and transforms nature, a process that started with the Neolithic and agricultural revolutions and continues today, for instance, through the application of genetic research. At the same time, scientists—and artists—construct and re-construct humanity's image of nature under the influence of historical and cultural patterns of thought.

This dynamic is reflected in the diversity of approaches taken by different schools of thought towards the analysis of human–nature relations. Social ecology studies societal relations to nature (Becker and Jahn 2006) with an explicit focus on the role of nature in supplying human needs (Hummel 2008; Lux 2009). Human ecology is a holistic 'interdiscipline' (Glaeser 2001), with roots in the early 20th century (Bews 1935; Park 1936). It frames its analysis around the interactions between humans, society and nature. Although society is seen as embedded within nature (Borden 2008), human ecology also interprets nature "as cultured by people" (Glaeser 1992). Environmental anthropologists see nature as part of different levels of social organization (Brondízio et al. 2009). Social-ecological systems (SES) analysis, with its origins in ecology (Holling 1973), takes society as embedded within nature as its point of departure. Thus, it took SES analysis longer than neighbouring fields in human–nature analysis to attract the attention of social scientists. Social science contributions to SES analysis mostly date from the late 1980s (Berkes 1989; Berkes et al. 1998; Blahna et al. 1998; Berkes et al. 2003). The points of departure for the most recent SES analyses are now located in ecology (Berkes and Folke 1998) as well as in the social sciences (Scheffer et al. 2003; Olsson et al. 2006; Armitage and Plummer 2010).

Individual schools of thought emphasise particular angles of the human–nature complex. The problem-oriented and spatial focus of governance and management–oriented SES analysis (see www.ecologyandsociety.org; Glaser, this volume; Wildenberg and Singh, this volume; Lopes and Begossi 2009) contrasts with the more theoretical and conceptual orientation of much work in human ecology (Glaeser 1989, 1995, 2003; Serbser 2004). The policy orientation of social ecology has its own distinctive emphasis on gender roles and human needs such as water, nutrition, mobility, housing, and associated supply systems (Becker 2004/Rev.2006; Becker and Jahn 2006).

While these varied approaches to the analysis of human–nature relations approach their inherent complexity in very different ways, they all study relationships, interactions or "couplings" between human and natural subsystems. These are, however, conceptualised in different ways, for example, as the study of societal metabolism (Fischer-Kowalski 2004), social-ecological feedbacks (Gunderson and Holling 2002), or societal relations with nature (Becker and Jahn 2006). The majority of analyses also concentrate at the local and regional level (Berkes et al. 2003; Anderies et al. 2006) rather than on planet Earth as a whole.

SES concepts function as "boundary objects" so as to facilitate communication and interaction between different disciplines, approaches and stakeholders (Becker, this volume). In this final chapter, we discuss how adjacent fields in human–nature analysis might move towards more integrated, collaborative research in a transdisciplinary complex systems mode. To this end, we review some relevant developments in the SES debate. We focus on what is needed to arrive at a problem-focused analysis that takes account of the planetary-level thresholds beyond which mainly anthropogenic environmental change endangers the "safe operating space" for humanity (Rockström et al. 2009a; Rockström et al. 2009b). Six questions relating to global sustainability research, which were developed from the original questions asked at the Sommerhausen Symposium (see Chapter 1), are discussed below.

WHAT IS REQUIRED FROM SUSTAINABILITY-ORIENTED SOCIAL-ECOLOGICAL ANALYSIS?

Methodological approaches should be subject to performance criteria. In our view, sustainability-oriented analyses of human–nature dynamics need to fulfil five minimum performance criteria, namely, to:

1. *Conform to the highest standards of academic practice*: SES analysis should be based on the transparent collection and rigorous interpretation of quantitative and qualitative data.
2. *Connect knowledge across diverse disciplines*: SES analysis should involve natural and social scientists in jointly designed, problem-

focused endeavours that integrate multiple forms of knowledge in collective learning processes. Results should contribute to the understanding of social and ecological capacities for self-organization and of emergent systemic properties such as resilience, vulnerability, and adaptive and transformative capacities.

3. *Link multi-level analysis across scales*: Local, national and regional analyses around the globe need to prepare the ground for integrated interdisciplinary larger-scale analyses of global change processes. The temporal and spatial scales of such multiple analyses need to be systematically linked across system levels.
4. *Facilitate participation and empowerment*: Analyses should enable ecosystem users and other influential stakeholders to exchange and reflect on their systemic knowledge and thus to increase their capacities to manage the natural systems to which their lives are linked.
5. *Link to policy-making, governance and management*: To facilitate the societal relevance of SES analyses, all relevant processes, from the planning of investigative ventures to the communication of scientific results, should contain an explicit two-way interface to relevant decision-makers.

WHO NEEDS COMPLEX SYSTEMS THINKING, AND WHAT FOR?

An SES can be seen as composed of multiple human and non-human agents whose interactions give rise to complex, non-linear dynamic properties (Glaser, this volume). In complexity theory, the notion of *complex* relates to the behaviour of a system rather than to its structure which implies a continual process of becoming or changing. Complex system behaviour is characterised by evolution, path dependency, non-linearity and emergence. Emergence arises from the interaction of the constituent elements of a system and is characterised by irreducible, counterintuitive, and unpredictable features (Ratter, this volume).

Who needs this type of systems thinking, which most people are not accustomed to? Which are the arenas where people can learn systems thinking and put knowledge in action? What tools, methods and approaches exist to bring systems thinking into society? These questions were debated during the World Systems Café exercise at the Sommerhausen Symposium (see Krause and Welp, this volume). A growing body of literature (Siebenhüner 2004; Keen et al. 2005; Pahl-Wostl et al. 2008; Renesch 2008; Gell-Mann 2010; Nobel Laureate Symposium 2011) agree that in order to identify and pursue more sustainable paths of global change, thinking in terms of integrated multi-level systems is needed by current and potential decision-makers throughout society, including children, ecosystem users, politicians, civil servants and global agencies. Systems thinking will increase the social understanding of human-driven bio-geophysical dynamics which, in turn, will contribute to the improvement of governance systems. In transdisciplinary research, systems thinking can be the interfacing language between different types of disciplinary knowledge, as well as between academic, ecosystem user, and other types of stakeholder knowledge. Systems thinking can thus play a crucial role in capacitating responsible action by increasing numbers of people at all levels of society and persuading democratic societies to shoulder the costs of the range of measures needed for the transition to sustainability (Speth 2008; Gell-Mann 2010).

Humanity needs systems thinking to better understand the ever-closer connections between human action and environmental change. Free flows of knowledge and information, in agreement with private intellectual property rights, are an essential precondition for the removal of the cognitive divides that arise from barriers to global communication. However, the public is in danger of being swamped by increasing volumes of bio-geophysical, social and economic data that are becoming available from global observation and monitoring systems. Systems thinking is necessary to generate an integrated view of ever more closely interwoven social-ecological change processes with this data (Steiner 2010; Sulston 2010) and to address the complexity of global social-ecological dynamics.

Any arena where learning takes places is a potential space for acquiring systems thinking capabilities: the family, the school and a wide range of other institutions dedicated to knowledge production, policy formulation and planning. The Internet and computer games can be used as tools for 'edutainment', that is, for learning in the context of fun and games. Face-to-face or virtual collaborative discussion and learning events, such as the World Systems café at the Sommerhausen Symposium, provide opportunities for interactive modelling, scenario building and visioning exercises, bringing brains together and shifting participants' values and priorities. Despite the influence of private vested interests, the media can play a particularly dynamic role in communicating and promoting the potential of systems thinking to an increasing number people worldwide (Krause and Welp, this volume).

Conventional schooling does not prepare us for understanding complex systems, with their non-linear behaviour, feedback loops and jumps in trajectories (Krause and Welp, this volume). To facilitate the emergence of a responsible, sustainability oriented world citizenry, strategies for the generation and effective transmission of systems-based knowledge are needed.

WHICH SOCIAL-ECOLOGICAL SYSTEM CONCEPTS BEST DESCRIBE WHICH TYPE OF SYSTEM?

Systems are the heuristic mental constructs of those who define them. This is not always explicitly acknowledged by system thinkers in the bio-geophysical sciences, who commonly take a positivist view of material phenomena, assuming that these can be "ground-truthed" in physical space. Indeed, it is a major preoccupation of the natural sciences to do just that, preferably using quantitative methods. Although constructivist approaches are not unknown (Stewart 2001), natural scientists mostly conceptualise, quantify and analyse living systems as real, material phenomena.

In contrast, social scientists conceptualise systems as composed of abstract, non-material elements such as institutions, values or communication (Luhmann 1984; Egner 2008). These are difficult to measure and more amenable to qualitative forms of analysis. Social systems are reflexive: Humans think about the systems they are part of, and even aggregate social agents collectively reflect on the implications of their behaviour and alter it as a result of such reflections (Young et al. 2006).

Thus constructivism comes more naturally to social scientists, while positivism is more plausible to natural scientists. In SES analysis, there is a creative tension between these two approaches to systems thinking. Humans are considered part of both bio-geophysical and social systems. Changes in knowledge, norms and values feed back into societal behaviour towards nature. Nature itself is regarded as more than a social construct; it is seen as endowed with a materiality and self-organisational forces of its own (Becker 1999).

The synthesis of positivist system concepts from the natural sciences which are supported by observation and experiment, with constructivist concepts from the social sciences which are supported by an analysis of human attitudes and behaviours, offers new potential for research on human–nature relations. A major challenge for this highly timely interdisciplinary venture is to develop procedures for defining and analysing SES dynamics that connect the physical space and properties of ecosystems to non-material social system features such as power and social cohesion.

Since humankind is an integral part and a powerful driver of Earth system dynamics (Nobel Laureate Symposium 2011) we can only observe these dynamics from the inside; at least at the planetary level, the observer is always part of the system. Moreover, complexity would appear to prohibit the development of comprehensive, universally applicable models of social-ecological dynamics. Thus social-ecological mind maps will always simplify in some sense or direction (Glaser 2006). This raises the question of how to decide which components to include in the concept or model of an SES—and which to leave out. These considerations introduce an explicitly constructivist element into any definition of an SES (Halliday and Glaser 2011). Driven by different underlying intentions, various paths towards analyzing co-evolving human–society–nature system dynamics have been taken.

To show how complex realities have been reduced to a variety of SES concepts, Table 10.1 presents a number of SES definitions alongside their special features and applications. Further examples are numerous—presented, for instance, by Cumming (2011: 38–40). In Table 10.1, the differentiating features of the SES definitions shown in Column 2 correspond to their respective purposes and uses in Column 3. Thus, the centrality of human needs and human resource uses emphasised in the Vienna and Frankfurt schools of social ecology (see first two SES definitions in Table 10.1) is well suited for steering the provisioning functions of nature into directions which are adequate to human needs. This, however, comes at the cost of a high degree of reductionism as far as bio-geophysical dynamics are concerned. Much of the SES work associated with the Stockholm school and the Resilience Alliance (www.resalliance.org), on the other hand, is rooted in an ecologically based understanding of resilience, adaptability and transformability in complex nested and adaptive systems. This advances long and medium-term analyses of co-evolving social and ecological systems at different scales, for example, as set out in the "general framework for analysing the sustainability of social-ecological systems" (Ostrom 2009) which assesses the likelihood of self-organisation among resource users. However, since social and ecological spheres are regarded as part of different disciplinary subsystems, scale mismatches between the social and ecological subsystems remain a problem (Cumming et al. 2006). In contrast, integrated SES definitions engage in a "parsing of social-ecological systems" that explicitly avoids the separation of social and ecological elements (Halliday and Glaser 2011) into separate subsystems. This approach eludes the scale mismatch problem between social and ecological spheres and is tailored towards practical social-ecological management in the growing number of areas where classic disciplinary divides are a hindrance.

The problem- or issue-oriented approach is a further path towards defining and modelling an SES. It is particularly appropriate for identifying the crucial leverage points for addressing specific pressing sustainability problems across the multiple levels of the global system whilst also remaining locally relevant (Glaser et al. 2010a). Problem-focused approaches lack the capacity for overarching system conceptualization, however. Unless appropriately embedded, they thus run the danger of shifting problems elsewhere in the global system rather than resolving them. The recently suggested concept of issue frameworks (Cumming 2011) as interlinked sets of problems around which to define SES, offers scope for circumnavigating this danger.

The final example in Table 10.1, Earth system analysis, addresses the (hitherto) highest possible spatial scale for SES analysis: the planetary level. While essential for the analysis of global threats, Earth system analysis has so far been less successful in translating global-level analyses into effective policy and governance measures. The subdivision of the planet into geo-, bio- and anthroposphere adopted by Earth system analysts may be at the root of this limitation. As identified in the currently ongoing global efforts to remedy this (ICSU 2010) integrated natural and social science analysis is still lacking in Earth system analysis.

The strength of SES analysis appears to lie in its diversity. While the limitations of each definition need to be explicitly acknowledged, agreement

Table 10.1 Differentiating Features and Applications of Some Social-Ecological Systems (SES) Definitions

SES Definition	Differentiating feature	Purpose/Application
Social Ecology: A boundary object to facilitate collaboration between scientific disciplines that needs to be developed into an epistemic object (Becker and Jahn 1999; Becker and Jahn 2006; Hummel 2008)	Nature is defined through its provisioning services to humanity (e.g. food and water)	To analyse those societal relations to nature which provide for human needs such as nutrition, water or mobility
Social Metabolism: The relation between humanity, nature and culture (Fischer-Kowalski 2004)	Society absorbs, transforms and emits flows of material and energy as it colonises nature	To support macro-economic calculations (e.g. the appropriation of net primary production)
Social-ecological systems: Integrated system of ecosystem and human society with reciprocal feedbacks and interdependence (Berkes and Folke 1998; Gunderson and Holling 2002; Olsson et al. 2004; Walker and Salt 2006; Ostrom 2009)	The adaptive renewal cycle and resilience. Different speeds and spatial scales are used to analyse adaptive change	To understand the source and role of economic, ecological and social change, particularly changes that are transforming, in systems that are adaptive.
Problem/issue focussed SES: A biogeophysical system with its associated social agents and institutions in a problem context (Glaser et al. 2010a)	Complexity is reduced by identifying the drivers of a problem at various levels of an SES	To analyse sustainability problems in specific realms, e.g. in coastal and marine systems or human settlements
Integrated SES : A system composed of organised assemblages of humans and non-human life forms in a spatially determined geophysical setting (Halliday and Glaser, 2011)	Four subsystems (Nature, World View, Control/Management and Technology) linked by exchanges of matter and meaning	To provide a generic system model as a tool for interdisciplinary social-ecological management
SES Governance: Governing system and system to be governed that are linked through governing interactions (Kooiman et al. 2005; Jentoft and Chuenpagdee 2009)	Relations between society and nature are conceptualised with an explicit governance focus	To understand interactions among natural, economic, social and governance systems in order to assess SES governability
Earth system analysis: The global SES encompasses the geosphere, biosphere and anthroposphere (Schellnhuber 1998; Schellnhuber et al. 2005; Schellnhuber et al. 2010)	An assembly of natural and social science models to construct the global picture	To understand the co-evolution of life and matter on Earth as one complex system to steer the planet safely through the Anthropocene

on a single definition of what an SES 'is' is therefore likely to reduce rather than strengthen analytical progress and relevance to global sustainability issues. Fitness for purpose rather than correspondence with 'reality' should be the principal criterion for judging the worth of any one of the variety of SES definitions and models.

HOW CAN SOCIAL-ECOLOGICAL SYSTEMS ANALYSIS CONTRIBUTE TO THE IDENTIFICATION OF FEASIBLE PATHWAYS TOWARDS SUSTAINABILITY?

Where large groups of diverse stakeholders might collaborate in social-ecological analysis, planning and implementation (Glaser and da Silva Oliveira 2004), transdisciplinary approaches which effectively link knowledge types are essential. However, methods for transdisciplinary social-ecological analysis are still in their infancy (Glaeser et al. 2009) Recent pioneer examples develop strategies to achieve coherence across large, multi-site, multi-disciplinary research programmes (the Coasts Under Stress Programme, Ommer 2007), employ frame-based modelling to prototype issue-based models with inputs from multiple disciplinary experts (Starfield and Jarre 2011) and with the example of 'ethno-oceanography' provide a blueprint on how to achieve systematic interfacing between a natural and a social science discipline around a specific issue (Gasalla and Diegues 2011; Starfield and Jarre 2011). Despite these encouraging forays into new interdisciplinary territory, knowledge creation is funded and careers continue to be made along predominantly disciplinary lines. Innovative methods for academically respected and societally effective global SES analysis need to be developed further to render us better able to explain complex SES dynamics at multiple levels, to support the development of empowering, transdisciplinary research practice which includes ecosystem users and other key decision-makers, and to generate new 'cutting edge' knowledge. The key challenge is to achieve synergies through collaborative methodological approaches.

Two major forces shape SES dynamics: spontaneous self-organisation and human-designed governance interventions. Governance initiatives will not necessarily have the intended effects: SES dynamics are non-linear policies; management are just one class of action taken by particular agents in the system. For instance, actions to reduce atmospheric CO_2 concentrations may be followed by the emergence of unintended and surprising system trajectories in the interplay between the multiple actions taken by a range of other agents. Since the behaviour of multiple agents is decisive for the history and future of a system, complex sustainability issues cannot be adequately addressed as simple processes. The key challenge is to understand *how* something complex can arise from simple behavioural rules for the individual agents (Ratter, this volume). Linear thinking in environmental policy can thus be inappropriate or even dangerous in a non-linear world (Krause and Welp, this volume). Innovative methods for SES analysis which take into account dynamic, non-linear social-ecological processes are needed to identify feasible pathways towards sustainability. Agent-based analysis of (inter)actions at the micro-level can then elucidate the often surprising phenomena that emerge at higher system levels.

Major agent types for SES analysis are individuals and aggregations of humans and non-human life forms (flora and fauna) while non-living entities (such as chemical particles) can also be conceptualised as agents. Agents differ in the scale of their actions and their properties: They could be fish in a reef, trees in a forest, people, households or larger societal organizations such as firms or nation states in a territory, or ecosystems in a biome. Unlike non-living agents, living agents differ in their behaviour. Behavioural differences between plants and animals reflect life histories and biological attributes such as genetic traits; those between humans are additionally associated with the mental maps, goals, abilities and endowments of the individuals concerned. The activities of individual agents typically give rise to patterns of dynamic behaviour associated with relatively stable emergent aggregates (Lansing and Kremer 1993). In SES, emergent aggregates can be social (e.g., the family, the firm, social networks), natural (e.g., forests, reefs) or social-ecological (fisheries, ecotourism and other resource management systems). These higher level aggregates have properties that cannot be explained directly from knowledge of the behaviour of the individual agents in the system; moreover they define an environment which influences the rules of interaction of these agents. To comprehensively understand a complex system, one has to analyse both top-down and bottom-up causality (see Ratter, this volume).

In the face of global sustainability concerns, the challenge for social-ecological research is to understand how contemporary societies and human–nature systems can evolve new patterns of aggregate behaviour that are well adapted to rapid global change and likely to promote the resilience of desirable SES features and the transformation of undesirable ones. Adaptive capacity is an increasingly important concept in social-ecological sustainability research and management. It has been defined as "the ability of a social-ecological system (or the components of that system) to be robust to disturbance and capable of responding to change" (Armitage and Plummer 2010). Adaptive capacity is a property of systems; however, it is rooted and reflected in the behaviour of individual agents.

Models are the principal tools for analyzing the adaptive capacity and patterns of behaviour of complex systems. Equation-based models focus on whole system behaviour under different scenarios, taking the internal workings of the system as given (Tretter and Halliday, this volume). Rule-based models, including agent-based modelling (Ferrand 2007), (re) construct complex system behaviour from the bottom up in accordance with the diverse behavioural rules employed by model agents. In agent-based models of SES, modifications of assumptions about the behaviour of natural and social actors lead to different emergent patterns and dynamics. While equation-based modelling is essential for the quantification of processes, the exploration of dynamic social-ecological processes which needs to precede quantification, including the analysis of emergence (Ratter, this volume) is better supported by agent-based approaches.

Agent-based models have been used in ecology and biology (artificial life), genetics, medicine, social sciences (artificial societies), economics, traffic analysis, physics, chemistry and art. They can combine knowledge from multiple disciplines in an integrated, spatially and temporally explicit simulation of system behaviour. In SES analysis, this potential is supported by transdisciplinary tools such as role-playing games, participatory modelling techniques (Glaser et al. 2010b) and interactive computer applications (Guyot and Honiden 2006). All these techniques ensure the participation of human system agents in social-ecological analysis (see CORMAS home page http://cormas.cirad.fr). Since human–nature dynamics are centrally affected by the perceptions, decisions and behaviour of human agents, social-ecological research must include the study of human stakeholders in the system. Stakeholder behaviour can be analysed in psychological terms and tested experimentally (Jager et al. 2000; Vollan and Ostrom 2010; Prediger et al. 2011). Stakeholder analysis can be combined with natural science assessments of bio-geophysical actions and reactions (Wildenberg and Singh, this volume) to address specific management and governance options at different temporal and spatial scales. The bottom-up, interdisciplinary reconstruction of emergent system behaviour through agent-based modelling can thus support the design of sustainability-oriented system interventions.

Integrated with laboratory and field experiments, surveys and participatory research, it can provide new insights into important as yet insufficiently understood variables, such as the influence of cultural differences on the emergence of rules-in-use for natural resource management (Poteete et al. 2010) or the impact of environmental and species characteristics on fishery institutions (Wilson et al. 2007). The development of transdisciplinary social-ecological models should build on the achievements of mono- and interdisciplinary beginnings and on the experiences with participatory modelling (Glaser, this volume). By allowing higher system levels to emerge from the interacting behaviours of social and natural agents, interdisciplinary agent-based modelling may even resolve the problem of spatial fit (Berkes et al. 1998; Galaz et al. 2008) between social and ecological scales.

WHAT ARE THE EXPERIENCES WITH STAKEHOLDER PARTICIPATION IN SOCIAL-ECOLOGICAL SYSTEMS ANALYSIS?

The drivers of social-ecological change originate at the local, regional or global level. In the Anthropocene, those drivers are most often associated with the behaviour of human individuals, social groups or organisations, that is, with specific groups of stakeholders. These engage in the maintenance, development or degradation of the complex social-ecological systems of which they are a part. Stakeholders self-organise (Ostrom 2009), contribute to and interact with environmental governance structures. Participation in social-ecological governance and management is a form of self-organised communication and a key driver of change in SES.

Participation can also be actively promoted and is the central component of a normative branch of social science and associated interdisciplinary research methods which aim to support and enable non-academic stakeholders in collaborative knowledge generation (Chambers 1992; Amartya and Loader 1999; Lebel and Bennet 2008). Participatory practice is similarly important for natural resource management, a field that has maintained a consistent focus on stakeholder participation over recent decades (Chambers 1997; Pahl-Wostl et al. 2007). In this sense, participation is emphasised in SES analysis to facilitate the management of system resilience and its central feature, self-organisational capacity, by stakeholders who are themselves part of the system (Walker et al. 2002).

To initiate and maintain participation in social-ecological analysis and action, an "emotional driver to reach people" (Krause and Welp, this volume) appears to be needed. This driver is often the perception of a resource crisis (Ostrom 2009). People feel emotional about what they value and care about and will engage when this appears threatened. A personal experience of the depletion of nature, but also hearing from friends or relatives, learning through the media or a piece of art, or simulating a crisis through role-play, are ways to activate those spheres of the mind which can increase participation in social-ecological management (Krause and Welp, this volume).

Over recent years, SES researchers have re-encountered a number of issues that have been faced by participatory rural appraisal and action researchers since the early 1980s. These include problems with the representativeness of participating stakeholders, inequalities of power and influence among stakeholders with opposing interests, including conflicts between the objectives and priorities of scientific knowledge generation and the interests of local stakeholders. The design of participatory tools for the analysis, planning and management of SES needs to respond to the characteristics of the ecological and social elements of the respective SES, and to the issues at hand. Thus, visual methods and the use of local materials and images may be crucial where the majority of participating stakeholders have little or no formal education. The written word, on the other hand, is an appropriate tool for participatory exercises with literate, formally educated groups that may react less readily to "local beans, sticks and stones" approaches (e.g., Chambers 1992).

Participatory research in democratic societies with a strong debating culture such as Britain may only need to offer platforms to spark off participation (Govan et al. 1998). In more explicitly hierarchical societies 'speaking up' may break prevailing societal norms of behaviour towards superiors, while livelihoods depend on being known as a loyal follower of a patron (Pelras 2000). In these contexts, participatory practice needs to tread a fine line between enabling the poorest and least powerful to participate meaningfully and endangering their livelihoods by inviting retributions from powerful patrons. This delicate balance can be pursued by 'nesting' participatory practice in the social structures of the respective environment, in ways which provide less powerful participants with additional room for manoeuvre for discussion and action without exposing their livelihood strategies to real or perceived dangers (Glaser et al. 2010b).

Widespread participation in decision-making can help SES in our changing world adapt to meet the challenges of sustainability. For instance, alternative development trajectories for coastal communities in many parts of the world are currently blocked because potential inputs from local knowledge are ignored by policy-makers (Krause et al. 2008; Glaeser and Glaser 2010). In the modelling realm, participatory methods can capacitate and empower local agents for self-organised management (Chambers 1994; Lebel et al. 2006). Where the construction and validation of agent-based models of human behaviour towards nature is undertaken in a participatory manner, this may challenge the cultural leanings of modellers by confronting model assumptions with the perceptions, values and priorities of system users (Glaser, this volume). In participatory agent-based modelling, SES analysis has a method which bridges two problematic divisions: that between the researcher and other SES stakeholders and, through an enhanced potential for the integration of natural science and social data into systems analysis, that between the social and natural sciences. An appropriately embedded participation of central stakeholders in SES analysis facilitates a fuller achievement of the social learning potential in SES and can thus improve sustainability prospects.

This participatory potential has so far been realised mainly at the local, regional and national levels (Lebel and Bennet 2008). At the global scale, the failure of the Climate Negotiations in 2009 in Copenhagen has led many people to question the transformational capacity of nation states and their ability achieve and implement agreements on the basis of intergovernmental agreements (Krause and Welp, this volume). Experiences like these raise questions about the capacity of democratic decision-making to deliver the substantial cuts in greenhouse gas emissions required to avoid catastrophic climate change and have led to calls for more autocratic decision-making. However, democracies have learning capacities that totalitarian systems do not possess (Held and Hervey 2009). These are rooted in participatory rights and practice and rely on features such as free public media, an independent scientific community, and the right to organise. Stakeholder participation in collaborative social learning platforms can provide important inputs for the understanding of SES dynamics and render systems thinking more widely accessible across socioeconomic divides. As the extent and quality of stakeholder participation increase in nested democratic resource management regimes, societal capacities for collaborative enquiry and innovative action are strengthened. This in turn promotes the adaptive and transformative capacities of resource governance and management systems (Berkes et al. 2005).

WHAT IS NEEDED TO VIEW THE EARTH AS A SYSTEM AND TO PRE-EMPT DANGEROUS CHANGE?

> Before I flew I was already aware of how small and vulnerable our planet is; but only when I saw it from space, in all its ineffable beauty and fragility, did I realize that human kind's most urgent task is to cherish and preserve it for future generations.
>
> —Sigmund Jähn, German Democratic Republic astronaut (cited on http://www.solarviews.com/germ/earthsp.htm accessed on August 13, 2010)

The first picture of our planet as a whole was taken from space on July 20, 1969. The public impact of this image promoted global awareness of planet Earth as a vulnerable entity. The early debate on Earth system sustainability was fuelled by the emotional impact of this positive image. More recently, the risks and uncertainties associated with growing global interconnectedness of human and non-human actors, technology and natural forces have been in the foreground. These concerns have been fuelled by specific events such as the Icelandic ash cloud in April 2010 and the Japanese earthquake/tsunami/ nuclear disaster in 2011 but also by an increasing consensus that ongoing Earth system changes such as climate change, biodiversity loss, nutrient loading and pollution are approaching globally critical thresholds. These delineate major biogeochemical boundaries which are now seen to circumscribe "the safe operating space for humanity" (Rockström et al. 2009a; Rockström et al. 2009b), although social thresholds such as human poverty and consumption levels have until recently remained largely outside the discussion. Beyond these boundaries, cascading social and ecological feedbacks across the globe endanger the ecosystem functions and services that sustain human life on Earth (Millennium Ecosystem Assessment 2005).

TRANSFORMATIONS

People usually become aware of the value of ecosystem services when these are already seriously impacted. Efforts have been made to attribute a monetary value to ecosystems and biodiversity on a global level (Costanza et al. 1997; Millennium Ecosystem Assessment 2005). The assumption behind such initiatives is that the problem is market failure with little consideration of intrinsic or spiritual value (Haggan 2011). How and by whom ecosystem services will be secured in future is a question of the right level of intervention and of effective instruments (Gómez-Baggethun et al. 2010). Consensus is also growing that to sustain ecosystem services and associated human well-being across the globe, a range of interlinked transformations in social and economic values, in demographic dynamics, in social equity, and in institutional structures and processes is needed (Raskin et al. 2002; Speth 2008; Gell-Mann 2010; Lucht 2010; Rockström et al. 2010). For instance, the German Advisory Council on Global Change stresses the need for a "social contract" for sustainability. This recognises that transformations in societies are not linear, and that the seeds of transformations may grow into something bigger in short periods of time.

A transformation is a fundamental alteration of the nature of a system which occurs—or becomes a societal objective—once current (ecological, social or economic) conditions become untenable or undesirable (Folke et al. 2009). Transformability is the ability of a system that is found undesirable, dysfunctional or subject to failure to successfully pursue desirable transformation (see Glaser et al., this volume: Box 1.1). The focus here is the feasibility and desirability of the system or system element in question. The need to transform undesirable systems is encapsulated in the idea of a transition towards sustainable development (Martens and Rotmans 2005).

At a local level, possible seeds of wider transformations can be seen, for example, in the transition town movement (Hopkins and Heinberg 2008). The opportunity to link activities in different cities, countries or even continents via the Internet plays a potentially crucial role. Learning from others' experiences and creating a shared identity of a transformation movement increase the speed and magnitude of such changes.

Global initiatives complement such bottom-up processes. In 2001, the newly formed Earth System Science Partnership called for Earth system management strategies. This led to the Earth System Governance Programme project that addresses

> the interrelated and increasingly integrated system of formal and informal rules, rule-making systems, and actor-networks at all levels of human society (from local to global) that are set up to steer societies towards preventing, mitigating, and adapting to global and local environmental change and, in particular, Earth system transformation, within the normative context of sustainable development. (Biermann et al. 2010)

The Earth System Governance Programme project is a long-term, interdisciplinary and social science–oriented global research venture which is active across a range of themes to understand the institutions and organisations through which humans govern their relationship to the natural environment (Biermann et al. 2010). However, an Earth system governance and management which achieves transformatory social-ecological change also requires a more integrated and thorough knowledge and understanding than we possess today of the way physical and biogeochemical Earth system dynamics are intertwined and co-evolve with human and societal behaviour. Humanity now needs not only to *conceive of* itself and its non-human environment as one system, but also to *understand* how this integrated, multi-dimensional and multi-level, local to global system works. Undesirable and destructive global phenomena like poverty and conflict need to be included in Earth system assessments and monitoring systems, and their feedbacks with bio-geochemical processes need to be assessed. This requires interdisciplinarity between the social and natural science as well as the location-specific integration of other forms of relevant knowledge in transdisciplinary approaches.

INTER- AND TRANSDISCIPLINARITY

Effective forms of generating and linking knowledge which include diverse disciplines and stakeholder groups across the Earth are needed for the global analysis of social-ecological sustainability. Until now, influential global analyses have occurred in separate spheres. World or generic type social systems and dynamics were the focus in the social sciences (Parsons 1951; Luhmann 1984; Wallerstein 2004) and planetary bio-geophysical dynamics in the natural sciences (Schellnhuber 1999; Schellnhuber et al. 2005). Despite recent attempts to develop integrated SES models that transcend the dichotomy between 'nature' and 'society' (Cumming 2011; Halliday and Glaser 2011), most contemporary SES definitions (see Table 10.1 and definitional options presented by Cumming 2011) still reflect the bifurcation between natural and social sciences.

Influential contemporary research into global environmental change has been similarly organised. In 2011, the International Geosphere-Biosphere Programme (IGBP) as the global umbrella for the natural sciences has about 20% social scientists including 'converted' natural scientists (www.igbp.net); the International Human Dimensions Programme on Global Environmental Change (IHDP) fulfils the same role for the social sciences (with about 10% natural scientists, www.ihdp.org). Among 12 core projects in the IGBP and IHDP groups, only the Land-Ocean Interactions in the Coastal Zone (LOICZ) and the Global Land Projects (GLP) operate under the joint IGBP/IHDP umbrella and are directed by scientific steering committees made up of natural and social scientists. This, separate natural and social science programmes and projects that operate in modular form with weak interfaces between natural and social scientists are still the rule in planetary-level social-ecological research.

A number of promising approaches that have been developed over recent decades point the way forward. Social and natural science analyses are clustering around a number of interdisciplinary system concepts such as sustainability, resilience, vulnerability, adaptation and transformational capacity. The definition of transformability, originally defined as relating only to the social realm (Carpenter et al. 2004), has been extended to include ecological features (e.g., Glaser et al., this volume: Box 1.1) so that the concept now features in social-ecological debates. Resilience, originally an ecological concept, has gained a prominent place in social and interdisciplinary science as "the capacity of a system to absorb disturbance, undergo change and still retain essentially the same function, structure, identity, and feedbacks" (see www.resalliance.org).

Inter- and transdisciplinary research programmes have been developed for supranational regions (e.g., the EU-funded SPICOSA, www.spicosa.eu) as well as subnational regions (e.g., the Swedish SUCOZOMA, www.mistra.org; the Brazilian MADAM, www.zmt-bremen.de/Binaries/Binary314/MADAM.pdf; and the Indonesian SPICE programmes, www.zmt-bremen.de/en/SPICE_II.html). Such programmes have further developed meaningful participatory practice in transdisciplinary research at the local and regional SES level (discussed earlier in this chapter). At the global level, an increasing range of stakeholders is also becoming involved in global priority setting and knowledge generation. Examples include the consultation on global sustainability research in which over 1000 participants from 85 countries commented and voted on the most urgent Earth system research questions (www.icsu-visioning.org/the-visioning-process; ICSU 2010) which served as the point of departure for an international scientific exercise on the "Grand Challenges to Global Sustainability Research" and the global consultation on the effects of climate change in which NGOs and research institutes involved over 4000 citizens from 38 countries in the production of an overview of global public opinion (www.wwviews.org).

Following concepts of co-evolutionary social-ecological dynamics (Norgaard 1994) and recognising growing global interconnectedness, recent global research initiatives are pursuing a more integrated analysis of coupled human–nature systems. The Earth System Science Partnership (ESSP, www.essp.org), in 2010 directed by 75% natural scientists and 25% social scientists (Rik Leemans, ESSP chair, pers. comm.), was formed from four largely disciplinary global environmental change programmes, which predominated in Earth system science until the turn of the Millennium: DIVERSITAS (www.diversitas-international.org), IGBP, IHDP and the World Climate Research Programme (WCRP, www.wcrp-climate.org). ESSP now operates four integrated problem/issue focused projects studying different aspects of Global Environmental Change (GEC): on the global Carbon Cycle (www.globalcarbonproject.org); GEC and food systems (www.gecafs.org); global water systems (www.gwsp.org); and, most recently, a fourth project on GEC and human health (www.gechh.org) (Leemans et al.

2009; Leemans et al. 2011). In the Stockholm Memorandum of May 2011 on "Tipping the Scales towards Sustainability", about 20 Nobel laureates and other renowned experts suggest priorities for global stewardship of the planet in the interest of future generations (www.globalsymposium2011.org). In this, science is called upon to generate knowledge which helps "to find solutions to the challenges of the Anthropocene . . . to launch a major initiative on the earth system and global sustainability . . . to tap all sources of ingenuity across disciplines and across the globe".

Different regions of the Earth are also organizing around Earth system sustainability. In Europe, the "Responses to Environmental and Societal Challenges for our Unstable Earth" (RESCUE) initiative of the European Science Foundations is developing new approaches towards interdisciplinary Earth system science. The African Network for Earth System Science (AfricanNESS) has completed a Science Plan (www.igbp.net). There is one regional research programme of the ESSP currently undertaking integrated social-ecological analysis: on human interaction with the monsoon system in Asia (see www.mairs-essp.org). While these are promising developments, the number of natural and social scientists who are actually involved with each other in integrated, collaborative global sustainability research exercises remains small in comparison to those still engaged in traditional mono-disciplinary research.

Not only is there still insufficient integration between the social and natural sciences in global sustainability analysis, but the social sciences are underrepresented, both in relative numbers (personal information Rik Leemans ESSP Chair, 16.08. 2010) and in the range of disciplines involved: This is likely to have contributed to the fact that contemporary global quantitative models and observation systems focus almost exclusively on bio-geophysical variables but include little economic and demographic data, while treating human cultures, values, norms and behaviour as external to the Earth system (Lucht 2010). Difficulties with interfacing quantitative with qualitative (see Tretter and Halliday, this volume), and spatially explicit regional with national data are another contributing factor to the imbalanced state of Earth system science practice.

Disciplinary imbalances in collaborative science affect the knowledge created. Most social scientists involved in the ongoing global analyses of the Earth System Science Partnership are geographers, economists, sociologists, political scientists and anthropologists. Behavioural scientists and historians, on the other hand, have to our knowledge, not contributed to planetary level collaborative analyses. Despite some relevant work on human behavioural types (Jager et al. 2000; Saunders 2003), analyses of the rationales and intentions underlying stakeholder behaviour towards ecosystems are still missing from present-day global sustainability analysis. Particularly needed are global analyses of the effects of bio-geophysical, social and economic factors on individual values and behaviour; on the connections between values and norms and human behaviour; and on the links between individual and group values and behaviour. Relevant disciplines include environmental psychology, which "deals with the relationship between the individual and his/her life space" (Moser and Uzzell 2004) and examines the temporal, social and physical variables that are relevant for social-ecological outcomes. Conservation psychology, a closely related field, investigates the motivations for conservation and for valuing the natural environment (Saunders 2003). Somewhat to the chagrin of some of their representatives, these and other psychological fields of enquiry have had little recognition by or impact on sustainability science so far (Gifford 2007). Given the known links between human health and environmental change (e.g. www.who.int/phe/en/; McMichael et al. 2008; Lara et al. 2011), further efforts in line with current beginnings under the Global Environmental Change and Human Health (GECHH) project are required to better integrate the health sciences into global sustainability research.

Representatives of such missing disciplines need to be involved from the start in designing new research ventures. Global nature observation systems and modelling endeavours need to also provide data on changes in cultures, societal values and human behaviour; and future Earth system models need to integrate such data.

One option, apparently not yet pursued, is to link global indices of human well-being, such as the United Nations Human Development Index (HDI) (see www.hdr.undp.org), with data from Earth systems research on natural dynamics. This was tried in the international scientific assessments on Climate (IPCC 2007), biodiversity, ecosystem services and human well-being (Millennium Ecosystem Assessment 2005) and agriculture (McIntyre et al. 2009), but success was limited because assessments relied only on the existing scientific literature. In future, teams of scientists with common points of departure and global aims such as the GLP, LOICZ and the ESSP Joint Projects will need to enrich existing data collection systems and scientific literature reviews with new integrative applications.

Ideally, then, collaborative strategies to generate and share Earth system sustainability knowledge and to nurture responsible Earth citizens would co-evolve with increasingly interdisciplinary data collection and analysis. Rifkin argues that the combination of new communication technologies with greater awareness about global environmental problems may create a new level of global consciousness and solidarity which reaches beyond the borders of nation states. For this, the systemic understanding of planetary level processes will need to enter culture (Rifkin 2009). This requires "well-founded narratives that are in full resonance with scientific findings" (Lucht and Pachauri 2004) so that "the collective mind of networked humanity" is better prepared to work on solutions (Lucht 2010). The engagement of systems scientists, the arts and humanities and the pedagogic sciences in this endeavour is needed.

In sum, to better understand planet Earth as an integrated system, it is necessary to connect the diverse knowledge systems that pre-dominate at the different levels of the global SES (Berkes 2011) and to widen and deepen interdisciplinarity.

EARTH SYSTEM SCALES AND LEVELS

Earth system dynamics evolve across multiple scales and levels. The differentiation of spatial and temporal as well as jurisdictional, institutional, governance and knowledge dimensions or scales of change is fairly recent (Cash et al. 2006; Leemans 2009). Each dimension/scale of change displays specific manifestations at different levels. Initiatives in multi-level, cross-scale analysis have set the stage for a deepening of research on human–nature interactions throughout the Earth system. A central sustainability-related question is, how do such multi-dimensional and cross-level interactions affect the resilience and transformability of the global SES? A widely accepted view of global institutional dynamics is that of Elinor Ostrom who views Earth system governance as a complex social system made up of multiple dynamic regimes nested within each other (Ostrom 1999). The idea of nested systems is common to a range of interdisciplinary social-ecological system concepts (Gunderson and Holling 2002; Gagnon and Berteaux 2009). However, mismatches or 'problems of fit' between ecological and institutional systems (Cumming et al. 2006; Ebbesson 2010) have obstructed effective interfacing between Earth system science, governance and management. Recent work distinguishes a range of 'misfit regimes' between bio-geophysical and resource management systems that need to be addressed to enable the effective pursuit of global sustainability objectives (Galaz et al. 2008).

The majority of models of social-ecological dynamics to date relate to a single spatial level: the local, regional or global (Leemans 2009). Global Earth system modelling avoids the abovementioned problems of fit by adopting an all-planet focus. It has generated important knowledge on feedbacks between atmosphere, hydrosphere, lithosphere and biosphere, and on thresholds and dangers at the global level. However, it has also largely fails to represent the divergent patterns and dynamics in different regions of the globe (Eisenack, this volume; Lucht 2010). Earth system analysis has thus had little resonance with national, regional and local decision-makers and its central conclusions have not been acted upon by policy-makers. This was highlighted in Copenhagen at the United Nation Climate Change Conference in December 2009 (COP15). In contrast, integrated social-ecological models that are constructed "from the bottom up" with close reference to local realities (e.g., Castella et al. 2005), simulating dynamics for specific localities or regions, successfully identify context-specific solutions that speak to local and regional policy-makers and are politically and ecologically sustainable (Cash et al. 2006). However, such bottom-up models are mostly local and regional and rarely link to global drivers and dynamics.

The challenge for global sustainability science now is to integrate multiple levels of analysis in multi-dimensional approaches (Lucht 2010).

Various upscaling methods from the local to the regional level have been used in agent-based modelling. A model of land use in Vietnam (Castella et al. 2005) uses historical analysis and a geographic information system to upscale findings from the level of the individual agent to that of the region; another (Wilson et al. 2007) derives the emergence of regional fisheries governance from the behavioural patterns of individual fishermen and target species characteristics and successfully uses regional landings data for ground-truthing. To address sustainability issues at the global level, innovative ways are now needed to interface such regional findings with our knowledge of global dynamics. A number of promising initiatives are pointing the way. Procedures have been developed that enable agent-based models to be linked to or embedded in other (equation-based) models (Kroeze et al. 2008). Wildenberg and Singh (this volume) adopt this approach to model tsunami-induced social transitions on the Nicobar Islands at the 'whole island' level. A number of authors are developing methods to assess the characteristics of social-ecological networks at higher system scales (Janssen et al. 2006; Okuyama and Holland 2008; Cinner and Bodin 2010; Fortuna et al. 2010; Smajgl 2010) and "metanetwork" analyses (Science—Special Section 2009) are being used to situate agent-focussed analyses in global dynamic frameworks.

Another path towards viewing the Earth as one system is the archetype approach (Eisenack, this volume). Archetypes are patterns of interaction between society and nature that bring about environmental change—or are brought about by it. The approach uses local and regional level data to identify complex relations and patterns that reappear in various parts of the Earth system. Archetypes provide the basis for a systematic classification of responses to social ecological problems in specific localities and regions across the globe. They enable comparative analysis and can be used to formulate context-specific policy recommendations while at the same time functioning as building blocks to construct a planetary level picture of human–nature dynamics from below. This intermediate level of abstraction and data resolution thus combines the strengths of top-down and bottom-up approaches to analysis and is a promising option for global sustainability analysis. Future research will need to assess whether the interfacing of social-ecological archetypes with the regionalisation of state and supranational level governance systems (Schulz et al. 2001) is possible and how this affects current "problems of fit" between ecological and institutional systems.

CONCLUSIONS

The articles in this book contribute to the generation of social-ecological knowledge, one of the preconditions for effective governance. Our contributions originate largely from the German-language academia, offered here within the necessarily global debate on human–nature interactions in the Anthropocene. We suggest that, in order to gain analytical depth and societal relevance, analyses of planetary changes need to address regionally diverse but integrated social-ecological dynamics within the wider analytical frame of a single integrated, multi-stakeholder Earth system.

FUTURE CHALLENGES

With increasing levels of societal colonisation of nature and other forms of human–nature connectivity, innovative approaches are needed to manage the impacts of environmental change on ecosystem services and human well-being (Fischer-Kowalski 2004; Clark et al. 2005; Millennium Ecosystem Assessment 2005; Lund and Benediktsson 2011). Transformations of both the structures and the knowledge generation practices in global sustainability research and Earth system governance are required to manage the diverse forms of global social-environmental change (Ellis et al. 2010; Glaeser and Glaser 2010; Krausmann and Fischer-Kowalski 2010). To increase the chances of pre-empting cascading crises by recognising critical thresholds and of effectively responding to unsustainable social-ecological trajectories, globally encompassing, adaptive frameworks for social-ecological sustainability research and for Earth system governance are required.

On the governance side, current legal systems are largely based on the principle of state sovereignty. This assigns political responsibility for the global commons and for transboundary ecosystems to nation states that are effectively free to, and do opt out of international agreements and cooperation (Ebbesson 2010). Alternative options exist: If the subsidiarity principle were applied at the planetary level, nested polycentric governance structures (Ostrom 2009) might operate to address specific social-environmental issues at those levels where problems are caused and suffered (Agrawal 2008). This may lay the groundwork for the development of more effective countervailing forces against pressing sustainability threats such as pollution and resource overuse than the current political steering mechanisms. Currently emerging transboundary river, coastal and coral reef regional management institutions point the way (Wilson et al. 2011). If endowed with inclusive and adaptive institutions and appropriately networked, a globally encompassing, polycentric approach to social-ecological governance could effectively connect sustainability-supporting sources of social energy across the globe (Glaser et al. 2003). For this to happen, the place-based searches for locally and regionally appropriate solutions which drive diverse regional, often culturally very specific initiatives (Berkes 2009; Pilgrim et al. 2010) need to be framed by a wider global visionary consensus on Earth system sustainability problems.

DIRECTIONS FOR SOCIAL-ECOLOGICAL RESEARCH

Current understandings of integrated social-ecological transformations are fragmented. Most contemporary SES definitions suggest that social and natural (sub)systems are separate but linked. As humanity has become the main driver behind environmental change including major planetary level threats in our new geological era of the Anthropocene, social-ecological analysis will need to reduce complexity in ways that more clearly represent the integrated social-ecological nature of Earth system dynamics. To increase the explanatory, prognostic and policy value of SES analysis, multi-level and multi-dimensional approaches are needed. It is time for an integrated, global social-ecological sustainability research paradigm that treats humanity and its well-being as an integral part of the Earth system, while fully including the context-specific character of local and regional dynamics.

A blueprint for a globally comprehensive, but regionally specific typology and assessment framework for SES appears required as a next step. The Millennium Ecosystem Assessment has made an important contribution to the development of a global framework for analysing human–nature interactions, principally in relation to the key role of ecosystem services in ensuring human well-being (Millennium Ecosystem Assessment 2005; Carpenter et al. 2009). Ellis and co-authors (Ellis et al. 2010) have classified and mapped human influence on the world's terrestrial ecosystems; their 'Antrome' (cf. biome) classification might be further refined by incorporating the human behaviour types associated with different institutional types and levels. Such initiatives prepare the ground for a conceptual framework to underpin a comprehensive supra-network of local and regional research networks. If associated with institutions which take on interstate, regional social-ecological governance functions (such as the EU, ASEAN or the CTI) such regionally grounded social-ecological research networks may well increase their societal and policy relevance.

A global, multi-phase, multi-stakeholder consultation was recently conducted to identify the research questions which might deliver to society "the knowledge and information necessary to assess the risks humanity is facing from global change and to understand how society can effectively mitigate dangerous changes and cope with the changes that we cannot manage" (ICSU 2010). Such inclusive processes are crucial for linking knowledge generation systems across disciplines, social groups and global regions, and for ensuring the relevance of global sustainability analysis to the place-specific priorities surrounding the governance of social and environmental change. Societal relevance is essential to generate the improved knowledge-action links that are needed to translate sustainability knowledge into sustainable practice. This also requires expansion of the bio-geophysical planetary boundaries concept (Rockström et al. 2009b) to incorporate social thresholds such as conflict, poverty, and inequality. Unless thresholds for variables such as poverty and conflict which are both a driver and an outcome of unsustainable human–nature relations are also included in Earth system analysis, this field's claim to "include the human occupants" (ICSU 2010) of planet Earth will ring hollow in the eyes of most people and policy-makers, on whose decisions the future of the planet depends. Global sustainability knowledge needs to be expanded and situated within an integrated system-based global research framework. This will require a reorganisation of science on the basis of "a coalition of scientists, scientific institutions and funding agencies that will commit to working together systematically" (ICSU 2010). We hope that this book is a contribution towards this pressing challenge.

ACKNOWLEDGMENTS

Thanks to Beate Ratter and Martin Welp for their comments on this chapter, to Kathleen Schwerdtner-Máñez and five anonymous reviewers for their detailed and insightful reviews, to Falk Schmidt and João Morais for providing information on the IDHP and IGBP programmes. Discussions with Rik Leemans updated us on the latest developments in Earth system science and helped to fine-tune our arguments.

REFERENCES

Agrawal, A (2008) *The Role of Local Institutions in Adaptation to Climate Change*. Report to the World Bank, Washington, DC

Amartya, L, Loader, R (1999) Participatory rural appraisal: Extending the research methods base. Agricultural Systems 62: 73–85

Anderies, JM, Walker, B, Kinzig, AP (2006) Fifteen weddings and a funeral: Case studies and resilience-based management. Ecology and Society 11: Art. 21 (online)

Armitage, D, Plummer, R (2010) *Adaptive Capacity and Environmental Governance Springer Series on Environmental Management*. Springer, Berlin, Heidelberg, Dordrecht, London, New York

Becker, E (1999) Fostering transdisciplinary research into sustainability in an age of globalization. In: Becker, E, Jahn, T (eds) *A Short Political Epilogue: Sustainability and the Social Sciences: A Cross-Disciplinary Approach Integrating Environmental Considerations into Theoretical Reorientation*. Zed, London, pp 284–289

Becker, E (2004/Rev.2006) Problem transformations in transdisciplinary research In: Hirsch Hadorn G (ed) *Unity of Knowledge in Transdisciplinary Research for Sustainability. Encyclopedia of Life Support Systems (EOLSS), Developed under the Auspices of the UNESCO*. Eolss Publishers, Oxford, http://eolss.net (accessed December 7, 2011)

Becker, E, Jahn, T (1999) *Sustainability and the Social Sciences: A Cross-Disciplinary Approach to Integrating Environmental Considerations into Theoretical Reorientation*. Zed, London

Becker, E, Jahn, T (2006) *Soziale Ökologie. Grundzüge einer Wissenschaft von den gesellschaftlichen Naturverhältnissen*. Campus, Frankfurt, New York

Berkes, F (1989) Cooperation from the perspective of human ecology. In: Berkes, F (ed) *Common Property Resources: Ecology and Community-Based Sustainable Development*. Belhaven Press. A division of Pinter Publishers, London, pp 70–88

Berkes, F (2009) Community conserved areas: Policy issues in historic and contemporary context. Conservation Letters 2: 19–24

Berkes, F (2011) Restoring unity: The concept of marine social-ecological systems. In: Ommer RE, Perry RI, Cochrane K, Cury P (eds) *World Fisheries: A Social-Ecological Analysis*. Wiley-Blackwell, Chichester, pp 9–28

Berkes, F, Colding, J, Folke, C (2003) *Navigating Social-Ecological Systems*. Cambridge University Press, Cambridge

Berkes, F, Colding, J, Folke, C (2005) *Navigating Social-Ecological Systems: Building Resilience for Complexity and Change*. Cambridge University Press, Cambridge, New York

Berkes, F, Colding, J, Folke, C, Pritchard, L, Jr., Svedin, U (1998) *The Problem of Fit Between Ecosystems and Institutions*. International Human Dimensions Programme (IHDP)

Berkes, F, Folke, C (1998) Linking social and ecological systems for resilience and sustainability. In: Berkes, F, Folke, C (eds) *Linking Social and Ecological Systems: Management Practices and Social Mechanisms for Building Resilience*. Cambridge University Press, Cambridge, pp 1–25

Bews, JW (1935) *Human Ecology*. Oxford University Press, London

Biermann, F, Betsill, MM, Vieira, SC, Gupta, J, Kanie, N, Lebel, L, Liverman, D, Schroeder, H, Siebenhüner, B, Yanda, PZ, Zondervan, R (2010) Navigating the anthropocene: The Earth system governance project strategy paper. Current Opinion in Environmental Sustainability 2: 202–208

Blahna, D, Brunson, MW, Endter-Wada, J, Krannich, RS (1998) A framework for understanding social science contributions to ecosystem management. Ecological Applications 8: 891–904

Borden, RJ (2008) A brief history of SHE: Reflections on the founding and first twenty five years of the Society for Human Ecology. Human Ecology Review 15: 95–108

Brondízio, ES, Ostrom, E, Young, OR (2009) Connectivity and the governance of multilevel social ecological systems: The role of social capital. Annual Review of Environmental and Resources 34: 253–278

Carpenter, SR, Holling, CS, Kinzig, A, Walker, B (2004) Resilience, adaptability and transformability in social-ecological systems. Ecology and Society 9: Art. 5 (online)

Carpenter, SR, Mooney, H, Agard, J, Capistrano, D, DeFries, RS, Diaz, S, Dietz, T, Duraiappah, AK, Oteng-Yeboah, A, Pereira, HM, Perrings, C, Reid, WV, Sarukhan, J, Scholes, RJ, Whyte, A (2009) Science for managing ecosystem services: Beyond the millenium ecosystem assessment. Proceedings of the National Academy of Sciences of the United States of America 106: 1305–1312

Cash, DW, Adger, W, Berkes, F, Garden, P, Lebel, L, Olsson, P, Pritchard, L, Young, O (2006) Scale and cross-scale dynamics: Governance and information in a multilevel world. Ecology and Society 11: Art. 8 (online)

Castella, JC, Trung, TN, Boissau, S (2005) Participatory simulation of land-use change in northern mountains of Vietnam: The combined use of an agent-based model, a role playing game, and a Geographic Information System. Ecology and Society 10: Art. 27 (online)

Chambers, R (1992) *Rural Appraisal: Rapid, Relaxed and Participatory*. Institute of Development Studies, Brighton

Chambers, R (1994) Participatory rural appraisal (PRA): Analysis of experience. World Development 22: 1253–1268

Chambers, R (1997) *Whose Reality Counts? Putting the First Last.* Intermediate Technology Publications, London
Cinner, J, Bodin, Ö (2010) Livelihood diversification in tropical coastal communities: A network-based approach to analyzing 'livelihood landscapes'. PlosONE 5: e11999
Clark, WC, Crutzen, PJ, Schellnhuber, HJ (2005) Abstract science for global sustainability: Toward a new paradigm. In: Schellnhuber, HJ, Crutzen, PJ, Clark, WC, Claussen, M, Held, H (eds) *Earth System Analysis for Sustainability.* MIT Press, Cambridge, MA, pp 1–28
Costanza, R, Cumberland, J, Daly, H, Goodland, R, Norgaard, RB (1997) An introduction to ecological economics. CRC Press LLC, Boca Raton
Cumming, GS (2011) *Spatial Resilience in Social-Ecological Systems.* Springer, New York
Cumming, GS, Cumming, DHM, Redman, CL (2006) Scale mismatches in social-ecological systems: Causes, consequences and solutions. Ecology and Society 11: Art. 14 (online)
Ebbesson, J (2010) The rule of law in governance of complex socio-ecological changes Global Environmental Change 20: 414–422
Egner, H (2008) *Gesellschaft, Mensch, Umwelt—beobachtet. Ein Beitrag zur Theorie der Geographie.* Steiner Verlag, Stuttgart
Ellis, EC, Klein Goldewijk, K, Siebert, S, Lightman, D, Ramankutty, N (2010) Anthropogenic transformation of the biomes, 1700 to 2000. Global Ecology and Biogeography 19: 589–606
Ferrand, N (2007) Modelling with and for stakeholders. In: Phan, D, Amblard, F (eds) *Agent-based Modelling and Simulation in the Social and Human Sciences.* The Bardwell Press, Oxford, pp 195–218
Fischer-Kowalski, M (2004) Gesellschaftliche Kolonisierung natürlicher Systeme: Arbeiten an einem Theorieversuch. In: Serbser, W (ed) *Humanökologie. Ursprünge-Trends-Zukünfte.* Oekom, München, pp 308–325
Folke, C, Chapin, FSI, Olsson, P (2009) Transformations in ecosystem stewardship. In: Chapin, FSI, Kofinas, GP, Folke, C (eds) *Principles of Ecosystem Stewardship: Resilience-Based Natural Resource Management in a Changing World.* Springer, New York, pp 103–125
Fortuna, MA, Stouffer, DB, Olesen, JM, Jordano, P, Mouillot, D, Krasnov, BR, Poulin, R, Bascompte, J (2010) Nestedness versus modularity in ecological networks: Two sides of the same coin? Journal of Animal Ecology 79: 811–817
Gagnon, CA, Berteaux, D (2009) Integrating traditional ecological knowledge and ecological science: A question of scale. Ecology and Society 14: Art. 19 (online)
Galaz, V, Olsson, P, Hahn, T, Folke, C, Svedin, U (2008) The problem of fit among biophysical systems, environmental and resource regimes, and broader governance systems: Insights and emerging challenges. In: Young, O, King, LA, Schroeder, H (eds) *Institutions and Environmental Change.* MIT Press, Cambridge, MA pp 147–186
Gasalla, MA, Diegues, ACS (2011) "Ethno-oceanography" as an interdisciplinary means to approach marine ecosystem change. In: Ommer, RE, Perry, RI, Cochrane, K, Cury, P (eds) *World Fisheries: A Social-Ecological Analysis.* Wiley-Blackwell, Chichester, pp 121–136
Gell-Mann, M (2010) Transformations of the twenty-first century: Transitions to greater sustainability. In: Schellnhuber, HJ, Molina, M, Stern, N, Huber, V, Kadner, S (eds) *Global Sustainability—a Nobel Cause.* Cambridge University Press, Cambridge, pp 1–8
Gifford, R (2007) Environmental psychology and sustainable development: Expansion, maturation and challenges. Journal of Social Issues 63: 199–212

Glaeser, B (1989) *Humanökologie—Grundlagen präventiver Umweltpolitik.* Westdeutscher Verlag, Opladen

Glaeser, B (1992) Natur in der Krise? In: Glaeser, B, Teherani-Kröner, P (eds) *Humanökologie und Kulturökologie: Grundlagen, Ansätze, Praxis.* Westdeutscher Verlag, Opladen, pp 49–70

Glaeser, B (1995) *Environment, Development, Agriculture: Integrated Policy Through Human Ecology.* University College London Press, London

Glaeser, B (2001) The changing human–nature relationship (HNR) in the context of GEC. In: Munn, T (ed) *The Encyclopedia of Global Environmental Change, Volume 5: Social and Economic Dimensions of Global Environmental Change (GEC)* John Wiley & Sons Ltd, Chichester, pp 11–24

Glaeser, B (2003) The changing human–nature relationships in the context of global environmental change. In: Timmerman, P (ed) *Encyclopaedia for Global Environmental Change. Volume 5: Social and Economic Dimensions of Global Environmental Change.* John Wiley, Chichester, pp 11–24

Glaeser, B, Bruckmeier, K, Glaser, M, Krause, G (2009) Social-ecological systems analysis in coastal and marine areas: A path toward integration of interdisciplinary knowledge. In: Begossi, A, Lopes, P (eds) *Current Trends in Human Ecology.* Cambridge Scholars Publishing, Cambridge, pp 183–203

Glaeser, B, Glaser, M (2010) Global change and coastal threats: The Indonesian case—an attempt in multi-level social-ecological research. Human Ecology Review—Special Issue 17: 135–147

Glaser, M (2006) The social dimension in ecosystem management: Strengths and weaknesses of human–nature mind maps. Human Ecology Review 13: 122–142

Glaser, M, Berger, U, Macedo, R (2003) Local vulnerability as an advantage: Mangrove forest management in Pará state, North Brazil under conditions of illegality. Regional Environmental Change 3: 162–172

Glaser, M, da Silva Oliveira, R (2004) The prospects for co-management of mangrove ecosystems on the North Brazilian coast—whose rights, whose duties and whose priorities? Natural Resources Forum 28: 224–233

Glaser, M, Krause, G, Oliveira, RS, Herazo-Fontalvo, M (2010a) Mangroves and people: A social-ecological system. In: Saint Paul, U, Schneider, H (eds) *Mangroves Dynamics and Management in North Brazil.* Springer, Heidelberg, Berlin, pp 307–354

Glaser, M, Radjawali, I, Ferse, S, Glaeser, B (2010b) 'Nested' community participation in hierarchical society? International Journal of Society Systems Science (online)

Gómez-Baggethun, E, de Groor, R, Lomas, PL, Montes, C (2010) The history of ecosystem services in economic theory and practice: From early notions to markets and payment schemes. Ecological Economics 69: 1209–1218

Govan, H, Harrison, M, Inglis, AS, Pretty, JN, Wightman, A (1998) Best practice in community participation for national parks. Scottish National Heritage Review 107

Gunderson, LH, Holling, CS (2002) *Panarchy. Understanding Transformations in Human and Natural Systems.* Island Press, Washington, DC, Covelo, London

Guyot, P, Honiden, S (2006) Agent-based participatory simulations: Merging multi-agent systems and role-playing games. Journal of Artificial Societies and Social Simulation 9 (online)

Haggan, N (2011) "You don't know what you've got 'til it's gone". In: Ommer RE, Perry RI, Cochrane K, Cury P (eds) *World Fisheries: A Social-Ecological Analysis.* Wiley-Blackwell, Chichester, pp 225–246

Halliday, A, Glaser, M (2011) A management perspective on social ecological systems: A generic system model and its application to a case study from Peru. Human Ecology Review 18 (1): 1–18

Held, D, Hervey, AF (2009) *Democracy, Climate Change and Global Governance: Democratic Agency and the Policy Menu Ahead.* Policy Network, London

Holling, CS (1973) Resilience and stability of ecological systems. Annual Review of Ecology and Systematics 4: 1–23

Hopkins, RJ, Heinberg, R (2008) *The Transition Handbook: From Oil Dependency to Local Resilience.* White River Junction, VT: Chelsea Green Publishing Company

Hummel, D (2008) *Population Dynamics and Supply Systems: A Transdisciplinary Approach.* Campus, Frankfurt, New York

ICSU (2010) *Earth Systems for Global Sustainability.* The Grand Challenges International Council for Science, Paris

IPCC (2007) *Climate Change 2007: Synthesis Report.* Intergovernmental Panel on Climate Change, Geneva

Jager, W, Janssen, MA, De Vries, HJM, De Greef, J, Vlek, CAJ (2000) Behaviour in commons dilemmas: Homo economicus and Homo psychologicus in an ecological-economic model. Ecological Economics 35: 357–379

Janssen, MA, Bodin, Ö, Anderies, JM, Elmqvist, T, Ernstson, H, McAllister, RRJ, Olsson, P, Ryan, P (2006) Towards a network perspective of the study of resilience in social-ecological systems. Ecology and Society 11: Art. 15 (online)

Keen, M, Brown, V, Dyball, R (2005) *Social Learning in Environmental Management: Towards a Sustainable Future.* Earthscan, London

Krause, G, Glaser, M, Lange, H (2008) Szenarien zur sozialen Gerechtigkeit an der Küste (Pará, Nordbrasilien). In: Bruckmeier, K, Serbser, W (eds) *Ethik und Umweltpolitik—Humanökologische Positionen und Perspektiven.* oekom Verlag, München, pp 324–348

Krausmann, F, Fischer-Kowalski, M (2010) *Gesellschaftliche Naturverhältnisse: Energiequellen und die globale Transformation des gesellschaftlichne Stoffwechsels.* Institute of Social Ecology, IFF—Faculty for Interdisciplinary Studies, Vienna

Kroeze, C, Middelburg, J, Leemans, R, Fennel, W, Glaser, M, Harashima, A, Liu, KK, Meybeck, M (2008) Integrating tools to assess change in semi-enclosed marine systems. In: Rizzoli, PM, Melillo, JM, Sundby, B, Urban, ER (eds) *Watersheds, Bays and Bounded Seas: The Science and Management of Semi-enclosed Marine Systems.* Island Press, Washington, DC, pp 77–96

Lansing, JS, Kremer, JN (1993) Emergent properties of Balinese water temple networks: Coadaptation on a rugged fitness landscape. American Anthropologist 95: 97–114

Lara, RJ, Neogi, SB, Islam, MS, Mahmud, ZH, Islam, S, Paul, D, Demoz, BB, Yamasaki, S, Nair, GB, Kattner, G (2011) Vibrio cholerae in waters of the Sunderban mangrove: Relationship with biogeochemical parameters and chitin in seston size fractions. Wetlands Ecology and Management 19: 109–119

Lebel, L, Anderies, J, Campbell, B, Hatfield-Dodds, S, Hughes, T, Wilson, J (2006) Governance and the capacity to manage resilience in regional social-ecological systems. Ecology and Society 11: Art. 19 (online)

Lebel, L, Bennet, E (2008) Participation in building scenarios of regional development. In: Norberg, J, Cumming, GS (eds) *Complexity Theory for a Sustainable Future.* Columbia University Press, New York, pp 209–222

Leemans, R (2009) Why regional and spatial specificity is needed in environmental assessments. In: Knight, G, Jäger, J (eds) *Integrated Regional Assessment of Global Climate Change.* Cambridge University Press, Cambridge, pp 312–331

Leemans, R, Asrar, G, Canadell, JG, Ingram, J, Larigauderie, A, Mooney, H, Nobre, C, Patwardhan, A, Rice, M, Schmidt, F, Seitzinger, S, Virji, H, Vörösmarthy, C, Young, O (2009) Developing a common strategy for integrative global change research and outreach: The Earth System Science Partnership (ESSP). Current Opinion in Environmental Sustainability 1: 4–13

Leemans, R, Rice, M, Henderson-Sellers, A, Noone, K (2011) Research agenda and policy input of the Earth System Science Partnership for Coping with Global Environmental Change. In: Brauch, HG, Oswald Spring, U, Mesjasz, C, Grin, J, Kameri-Mbote, P, Chourou, B, Dunay, P, Birkmann, J (eds) *Coping with Global Environmental Change, Disasters and Security: Threats, Challenges, Vulnerabilities and Risks*. Springer, Berlin, pp 1205–1220

Lopes, P, Begossi, A (2009) *Current Trends in Human Ecology*. Cambridge Scholars Publishing, Newcastle

Lucht, W (2010) Earth system analysis and taking a crude look at the whole. In: Schellnhuber, HJ, Molina, M, Stern, N, Huber, V, Kadner, S (eds) *Global Sustainability: A Nobel Cause*. Cambridge University Press, Cambridge, pp 19–32

Lucht, W, Pachauri, RK (2004) The mental component of the Earth system. In: Schellnhuber, HJ, Crutzen, PJ, Clark, WC, Claussen, M, Held, H (eds) *Earth System Analysis for Sustainability*. Cambridge University Press, Cambridge, pp 341–365

Luhmann, N (1984) *Soziale Systeme. Grundriss einer allgemeinen Theorie*. Suhrkamp, Frankfurt am Main

Lund, KA, Benediktsson, K (2011) Inhabiting a risky Earth: The Eyjafjallajökull eruption in 2010 and its impacts. Anthropology Today 27: 6–9

Lux, A (2009) *Wasserversorgung im Umbruch. Der Bevölkerungsrückgang und seine Folgen für die öffentliche Wasserwirtschaft Campus Forschung, Bd. 398*. Campus Publishers, Frankfurt am Main, New York

Martens, P, Rotmans, J (2005) Transitions in a globalising world. Futures 37: 1133–1144

McIntyre, BD, Herren, HR, Wakhungu, J, Watson, RT (2009) *International Assessment of Agricultural Knowledge, Science and Technology for Development (IAASTD): A Synthesis of the Global and Sub-Global IAASTD Reports*. Island Press, Washington, DC

McMichael, AJ, Friel, S, Nyong, A, Corvalan, C (2008) Global environmental change and health: Impacts, inequalities, and the health sector. British Medical Journal 336: 191–194

Millennium Ecosystem Assessment (2005) *Ecosystems and Human Well-being: Synthesis*. Island Press, Washington D. C.

Moser, G, Uzzell, D (2004) Psychology and the challenge of global environmental change. IHDP Update. Newsletter of the international Human Dimensions Programme on the Gobal Environmental Change 4: 1–2

Nobel Laureate Symposium (2011) *The Stockholm Memorandum: Tipping the Scales Towards Sustainability*. 3rd Nobel Laureate Symposium on Global Sustainability, 16–19 May 2011, Stockholm

Norgaard, RB (1994) *Development Betrayed: The End of Progress and a Coevoluntionary Revisioning of the Future*. Routledge, London, New York

Okuyama, T, Holland, JN (2008) Network structural properties mediate the stability of mutualistic communities. Ecology Letters 11: 208–216

Olsson, P, Gunderson, LH, Carpenter, SR, Ryan, P, Lebel, L, Folke, C, Holling, CS (2006) Shooting the rapids: Navigating transitions to adaptive governance of social-ecological systems. Ecology and Society 11: Art. 18 (online)

Ommer, RE (2007) *Coasts Under Stress: Restructuring and Social-Ecological Health*. McGill-Queens University Press, Montreal & Kingston, London, Ithaca

Ostrom, E (1999) Coping with tragedies of the commons. Annual Review of Political Science 2: 493–535

Ostrom, E (2009) *A Polycentric Approach for Coping with Climate Change*. Background Paper to the 2010 World Development Report. Policy Research Working Paper 5095. World Bank, Washington D. C.

Pahl-Wostl, C, Mostert, E, Tàbara, D (2008) The growing importance of social learning in water resources management and sustainability science. Ecology and Society 13: Art. 24 (online)
Pahl-Wostl, C, Sendzimir, J, Jeffrey, P, Aerts, J, Berkamp, G, Cross, K (2007) Managing change toward adaptive water management through social learning. Ecology and Society 12: Art. 30 (online)
Park, RE (1936) Human ecology. In: Hughes, EC, Johnson, CS, Masuoka, J, Redfield, R, Wirth, L (eds) Reprinted in: *Human Communities: The Collected Papers of Robert Ezra Park, Vol. II* (1952) The Free Press of Glencoe, Glencoe, pp 145–158
Parsons, T (1951) *The Social System.* The Free Press, New York
Pelras, C (2000) Patron-client ties among the Bugis and Makassarese of South Sulawesi. Bijdragen tot de Taal-; Land- en Volkenkunde 156: 39
Pilgrim, S, Samson, C, Pretty, J (2010) EcoCultural revitalisation: Replenishing community connections to the land. In: Pilgrim S, Pretty J (eds) *Nature and Culture: Rebuilding Lost Connections.* Earthscan, London, pp 235–256
Poteete, AR, Janssen, MA, Ostrom, E (2010) *Working Together: Collective Action, the Commons, and Multiple Methods in Practice.* Princeton University Press, Princeton, NJ
Prediger, S, Vollan, B, Frölich, M (2011) The Impact of Culture and Ecology on Cooperation in a Common-pool Resource Experiment. Ecological Economics 70 (9): 1559–1608
Raskin, PD, Banuri, T, Gallopin, G, Gutman, P, Hammond, A, Kates, RW, Swart, R (2002) *Great Transition: The Promise and Lure of the Times Ahead.* Stockholm Environment Institute, Boston
Renesch, J (2008) A systems view of corporate culture change. The Ecologist, 11 June, first appeard in the Ecologist in February (online)
Rifkin, J (2009) *The Empathic Civilization: The Race to Global Consciousness in a World in Crisis.* Jeremy P. Tarcher, Penguin Group USA, New York
Rockström, J, Steffen, W, Noone, K, Persson, A, Chapin, FS, Lambin, EF, Lenton, TM, Scheffer, M, Fole, C, Schellnhuber, HJ, Nykvist, B, de Wit, CA, Hughes, T, van der Leeuw, S, Rodhe, H, Sörlin, S, Snyder, PK, Constanza, R, Svedin, U, Falkenmark, M, Karlberg, L, Corell, RW, Fabry, VJ, Hansen, J, Walker, B, Liverman, D, Richardson, K, Crutzen, P, Foley, JA (2009a) A safe operating space for humanity. Nature 461: 472–475
Rockström, J, Steffen, W, Noone, K, Persson, A, Chapin, FS, Lambin, EF, Lenton, TM, Scheffer, M, Folke, C, Schellnhuber, HJ, Nykvist, B, de Wit, CA, Hughes, T, van der Leeuw, S, Rodhe, H, Sörlin, S, Snyder, PK, Constanza, R, Svedin, U, Falkenmark, M, Karlberg, L, Corell, RW, Fabry, VJ, Hansen, J, Walker, B, Liverman, D, Richardson, P, Crutzen, P, Foley, J (2009b) Planetary boundaries: Exploring the safe operating space for humanity. Ecology and Society 14: Art. 32 (online)
Rockström, J, Vohland, K, Lucht, W, Lotze-Campen, H, von Weizsäcker, EU, Banuri, T (2010) Making Progress within and beyond borders. In: Schellnhuber, HJ, Molina, M, Stern, N, Huber, V, Kadner, S (eds) *Global Sustainability: A Nobel Cause.* Cambridge University Press, Cambridge, pp 33–47
Saunders, C (2003) The emerging field of conservation psychology. Human Ecology Forum 10: 137–149
Scheffer, M, Szabo, S, Gragnani, A, van Nes, EH, Sergio Rinaldi, S, Kautsky, N, Norberg, J, Roijackers, RMM, Franken, RJM (2003) Floating plant dominance as a stable state. PNAS 100: 4040–4045
Schellnhuber, HJ (1999) 'Earth system' analysis and the second Copernican revolution. Nature 402: 19–23
Schellnhuber, HJ, Crutzen, PJ, Clark, WC, Hunt, J (2005) Earth system analysis for sustainability. Environment 47: 11–25

Schulz, M, Söderbaum, F, Öjendal, J (2001) *Regionalizing in a Globalizing World: A Comparative Perspective on Forms, Actors and Processes.* Zed Books Ltd., London, New York

Science—Special Section (2009) Complex systems and networks. Science 325

Serbser, W (2004) *Humanökologie. Ursprünge—Trends—Zukünfte.* Oekom Verlag, München

Siebenhüner, B (2004) Social learning and sustainability science: Which role can stakeholder participation play? International Journal of Sustainable Development & World Ecology 7: 146–163

Smajgl, A (2010) Challenging beliefs through multi-level participatory modelling in Indonesia. Environmental Modelling & Software 25: 1470–1476

Speth, JG (2008) *The Bridge at the End of the World: Capitalism, the Environment and Crossing from Crisis to Sustainability.* Yale University Press, New Haven, CT

Starfield, AM, Jarre, A (2011) Interdisciplinary modeling for an ecosystem approach to management in marine social-ecological systems. In: Ommer, RE, Perry, A, Cochrane, K, Cury, P (eds) *World Fisheries: A Social-Ecological Analysis.* Wiley-Blackwell, Chichester, pp 105–119

Steiner, A (2010) Democracy and participation. In: Schellnhuber, HJ, Molina, M, Stern, N, Huber, V, Kadner, S (eds) *Global Sustainability: A Nobel Cause.* Cambridge University Press, Cambridge, pp 361–366

Stewart, J (2001) Radical constructivism in biology and cognitive science. Foundation of Science 6: 99–124

Sulston, J (2010) Information flow: The basis for sustainable participation. In: Schellnhuber, HJ, Molina, M, Stern, N, Huber, V, Kadner, S (eds) *Global Sustainability: A Nobel Cause.* Cambridge University Press, Cambridge, pp 343–354

Vollan, B, Ostrom, E (2010) Cooperation and the commons. Science 330: 923–924

Walker, B, Carpenter, S, Anderies, JM, Abel, N, Cumming, GS, Janssen, MA, Lebel, L, Norberg, J, Peterson, G, Pritchard, R (2002) Resilience management in social-ecological systems: A working hypothesis for a participatory approach. Conservation Ecology 6: Art. 14 (online)

Wallerstein, I (2004) *World-Systems Analysis: An Introduction.* Duke University Press, Durham, NC

Wilson, J, Darmawan, A, Subijanto, J, Green, A, Sheppard, S (2011) *Scientific Design of a Resilient Network of Marine Protected Areas. Lesser Sunda Ecoregion, Coral Triangle.* Report 2/11. Asia Pacific Marine Program

Wilson, J, Yan, L, Wilson, C (2007) The precursors of governance in the Maine lobster fishery. PNAS 104: 15212–15217

Young, OR, Berkhout, F, Gallopin, GC, Janssen, MA, Ostrom, E, van der Leeuw, S (2006) The globalization of socio-ecological systems: An agenda for scientific research. Global Environmental Change 16: 304–316

Contributors

Egon Becker
Institute for Social-Ecological Research (ISOE)
becker@isoe.de

Prof. Dr. Egon Becker is a retired professor of theory of science and sociology of higher education at the Goethe-University Frankfurt/Main (Germany) and senior researcher at the Institute for Social-Ecological Research (ISOE). He studied mathematics and physics, sociology and philosophy, graduated in electrical engineering and physics, and has a doctoral degree in theoretical physics in the field of quantum theory of solid states from the Technical University Darmstadt, Germany. He was research fellow at Yale University (US) and visiting professor in Brazil, Mexico and Sweden. Currently his fields of research are conceptual and methodological problems of social-ecological research on which he has published extensively. Recent publications include *Border Zones of Ecology and Systems Theory* (2011) (with Broder Breckling) and *Gesellschaftliche Naturverhältnisse als Rahmenkonzept* (2011) (with Diana Hummel and Thomas Jahn).

Klaus Eisenack
Carl von Ossietzky Universität Oldenburg
klaus.eisenack@uni-oldenburg.de

Prof. Dr. Klaus Eisenack is assistant professor for environment and development economics at Carl von Ossietzky University Oldenburg, Germany. He worked at the Potsdam Institute for Climate Impact Research from 2001 to 2008. His current research focuses on adaptation to climate change. He is leader of the Chameleon research group that investigates adaptation of public and private utilities. In 2006 he received a PhD in mathematics at the Free University Berlin with a work on qualitative and non-standard modelling techniques with applications to natural resource management. Further interests are modelling of policy instruments, game theory, archetypical patterns of social-ecological systems and games on global change.

Bernhard Glaeser
German Society for Human Ecology (DGH), Free University Berlin (FUB)
BernhardGlaeser@gmail.com

Prof. Dr. Bernhard Glaeser is the president of the German Society for Human Ecology (DGH), Land-Ocean Interactions in the Coastal Zone (LOICZ) corresponding member, and member of Integrated Marine Biogeochemistry and Ecosystem Research (IMBER) Human Dimensions. He joined the Social Science Research Center Berlin (WZB) in 1976, became professor of human ecology in 1995 at Göteborg University (Sweden) and professor of sociology in 2001 at the Berlin Free University, Department of Political and Social Sciences. Glaeser holds degrees in economics, philosophy and sociology. He has done environmental and sustainability research in Germany, Sweden, Tanzania, China, India and Indonesia and is affiliated with various international editorial and advisory boards.

Marion Glaser
Leibniz-Center for Tropical Marine Ecology
Marion.Glaser@zmt-bremen.de

PD Dr. Marion Glaser is an environmental sociologist at the Leibniz Center for Tropical Marine Ecology (ZMT), Bremen, Germany. She trained at the Universities of Cologne, Bath and London and worked in the UK, Bangladesh and Belize on agriculture, forestry, flood control and coastal management. She led the social science research in the Brazilian Mangrove Dynamics and Management (MADAM, 1996–2005) and the Science for the Protection of Indonesian Coastal Ecosystems (SPICE, 2007–2010) programmes. Today, she is Executive Board member of the Land-Ocean Interactions in the Coastal Zone (LOICZ) programme, vice-president of the German Society for Human Ecology and leads the Social-Ecological Systems Group at the ZMT. Her work focuses on transdisciplinary sustainability analyses of tropical coastal and marine social-ecological systems and she has (co)authored over 50 articles and books.

Andrew Halliday
andrewhalliday2001@yahoo.co.uk

Andrew Halliday worked as a manager and consultant for integrated conservation and development projects in Latin America, specialising in participatory monitoring and evaluation. He is currently a freelance academic editor and teaches environment, development and international studies as an associate lecturer at the Open University, UK, while pursuing his research interest in social-ecological systems.

Diana Hummel
Institute for Social-Ecological Research (ISOE)
hummel@isoe.de

PD Dr. Diana Hummel is political scientist at the Institute for Social-Ecological Research (ISOE), Frankfurt/Main, Germany and lecturer at the Institute of Political Science, Goethe-University. She trained political science, sociology and educational science at the Goethe-University of Frankfurt. Today, she is senior scientist and head of the research area "Population Development and Supply" at ISOE and is engaged in transdisciplinary analyses on social-ecological supply systems based on natural resources. Key publications include Hummel, D (2011) Bevölkerungsentwicklung, Ökologie und Versorgung. In: Groß, M (ed): *Handbuch Umweltsoziologie*. Wiesbaden, pp 565–597; Hummel, D (ed) (2008) *Population Dynamics and Supply Systems: A Transdisciplinary Approach*. Frankfurt/New York.

Gesche Krause
Leibniz-Center for Tropical Marine Ecology
gesche.krause zmt-bremen.de
Dr. Gesche Krause is a coastal geographer at the Leibniz Center for Tropical Marine Ecology (ZMT), Bremen, Germany. She trained at the Universities of Kiel, Greifswald (Germany), and Stockholm (Sweden) and worked in the UK and US. She led the geographical science research in the Brazilian Mangrove Dynamics and Management (MADAM, 1996–2005) and is involved in the Science for the Protection of Indonesian Coastal Ecosystems (SPICE) programme on ornamental trade dynamics and coastal governance issues. She is co-founder of the ZMT Social-Ecological Systems (SES) Analysis research group and leader of the ICES study group on social dimensions of aquaculture.

Rik Leemans
Wageningen UR
rik.leemans@wur.nl
Prof. Dr. Rik Leemans directs the Environmental Systems Analysis group and the WIMEK graduate school at Wageningen University. He leads several international projects on global change and sustainability. He contributed to the Intergovernmental Panel on Climate Change (IPCC) and the Millennium Ecosystem Assessment. His early research emphasised modelling boreal forest dynamics. Later, he led the development and application of an integrated earth-system model, IMAGE. His current research concerns biogeochemical cycles, biodiversity and ecosystem services, and sustainable development. He published many papers on a wide range of environmental topics and is editor-in-chief of the journal *Current Opinion in Environmental Sustainability*.

Beate M.W. Ratter
Universität Hamburg
Institut für Geographie
ratter@geowiss.uni-hamburg.de
Prof. Dr. Beate M.W. Ratter is professor for integrative geography at the Institute of Geography, University of Hamburg and holds a joint

position as head of the Department of Human Dimension of Coastal Areas at the Institute of Coastal Research, Helmholtz Centre Geesthacht. Her research areas are the analysis of resource and environmental management strategies including perception in different regional cultures. Current research interests focus regional planning and climate change. Regional research expertise was gained along the European North Sea coast, in the Caribbean, Latin America, Canada and Taiwan. From 2000 to 2002 she was project leader for the "Trilateral Wadden Sea" WWF-International, Bremen. From 2002 to 2007 she was professor of human geography, Johannes Gutenberg-University Mainz. She published and authored over 80 articles and books.

Simron Jit Singh
Institute of Social Ecology
IFF—Fakultät für interdisziplinäre Forschung und Fortbildung der
Alpen-Adria-Universität Klagenfurt
simron.singh@uni-klu.ac.at
Dr. Simron Jit Singh (Ph.D., human ecology) is a senior researcher at the Institute of Social Ecology Vienna. His research focuses on the theoretical, conceptual and empirical aspects of society-nature interactions within the framework of sustainability science and the development discourse. He has undertaken extensive fieldwork among the pastoral nomadic community, the Van Gujjars of the Central Indian Himalayas and the Nicobarese of the Nicobar Islands, India. He is recipient of several awards and fellowship and has published extensively, including books. He teaches ecological anthropology, methods in local studies and international development at Austrian universities.

Felix Tretter
Isar-Amper-Clinics
felix.tretter@iak-kmo.de
Prof. Dr. Felix Tretter is a senior physician at the Isar Amper Clinic in Munich leading the department for addiction. He is also a teacher for psychology at the University of Munich. He studied medicine, social science psychology and philosophy and has been involved in human ecology and systems science since 1975. He is interested in theoretical human ecology. He also was working in experimental neurobiology. At present, he is involved in philosophical anthropology and computational systems biology of mental disorders. He (co-) authored about 120 articles and books.

Martin Welp
Eberswalde University for Sustainable Development
Faculty of Forest and Environment
martin.welp@hnee.de
Prof. Dr. Martin Welp is professor for socioeconomics and communication at the Eberswalde University for Sustainable Development. He

is head of the International Master Study Programme Global Change Management. Before his appointment in Eberswalde he worked as senior researcher at the Potsdam-Institute for Climate Impact Research (PIK) and at the Technical University Darmstadt. He has headed research projects on science-based stakeholder dialogues on climate mitigation and adaptation, as well as public participation in natural resources management. Projects he is currently involved in deal with climate adaptation in the field of (urban) forestry and water resources management in Northwest China (Tarim River).

Martin Wildenberg
GLOBAL 2000/Friends of the Earth Austria,
martin.wildenberg@global2000.at

Martin Wildenberg works at the Austrian environmental NGO GLOBAL 2000 (Friends of the Earth Austria) on the topic of sustainable agricultural production. He is also a PhD research fellow at the Institute of Social-Ecology in Vienna. His focus is on modelling local social-ecological systems under transition using agent-based and soft-system modelling tools. He has developed the first freely available software for Fuzzy Cognitive Mapping, which can be downloaded at www.fcmappers.net. He was involved in ALTER-Net where he mainly worked on an agent-based land-use change model of three European LTSER sites and the evaluation of fuzzy cognitive mapping for its usability in integrated conservation research and management. Beside the work for ALTER-Net he has also worked in a project concerned with the development and rebuilding of the Nicobar Islands after the 2004 Tsunami and a participative agent-based modelling project, focusing on a rural Austrian municipality. He is especially interested in using participative modelling and mapping approaches in applied settings and in understanding the complex dynamics of change driving local social-ecological system.

Index